Decoding Evolution

Volume 1

Series Editor

Carlo Pecoraro, Scientific Director, Physalia Courses, Berlin, Germany

The textbook series *Decoding Evolution* is based on the editor's experience teaching undergraduates and assisting researchers who need to perform their own bioinformatic analyses. Each volume highlights the complexities of applying bioinformatics methods to real evolutionary questions and helps researchers to overcome these challenges in their daily work. They also promote the transfer of new methods and emerging techniques to a wide range of readers, from undergraduate students to university educators and scientists, addressing the growing need to learn new skills to accommodate and analyze large data sets.

Full-color illustrations, step-by-step protocols, and straightforward take-home messages provide an accessible approach to learning and understanding.

All volumes in this series are based on lectures from *Physalia* -courses (https://www.physalia-courses.org/), a company that provides scientific training courses and workshops in bioinformatics, genomics, and related fields.

Filippo Biscarini • Nelson Nazzicari

Deep Learning for Life Sciences

with Python Notebooks for Examples and Exercises

Filippo Biscarini
National Research Council CNR
Milano, Italy

Nelson Nazzicari
Research Centre for Animal Production and Aquaculture
Council for Agricultural Research and Analysis of Agricultural Economics
Lodi, Italy

ISSN 2662-7442 ISSN 2662-7450 (electronic)
Decoding Evolution
ISBN 978-3-031-96851-8 ISBN 978-3-031-96852-5 (eBook)
https://doi.org/10.1007/978-3-031-96852-5

This Springer imprint is published by the registered company Springer Nature Switzerland AG
The registered company address is: Gewerbestrasse 11, 6330 Cham, Switzerland

If you can't explain something in simple terms,
you don't understand it.

—Richard Feynamn

Editor's Note

This book kicks off the "Decoding Evolution" series, a collaboration between Physalia-courses (www.physalia-courses.org) and Springer Nature. The series aims to demystify bioinformatics, statistics, and data science, making these topics approachable even for readers new to the field. Filippo and Nelson have nailed it with this book. They break down complex concepts like Deep Learning into practical, easy-to-understand language, inspiring readers to dive into using these models on their own data, without the usual fear and hesitation that can hold beginners back.

Contents

Part II Neural Networks and Deep Learning Decoded

Part III Making It Work

List of Figures

List of Tables

Introduction

1

1.1 Why a Book on Deep Learning for Life Sciences?

Before telling you why you should read this book, we think it's fair to spend a few words explaining why we wrote it in the first place. Since 2020, we have been running a successful course on deep learning, now—in 2025—in its 6th edition ("Introduction to Deep Learning", at Physalia-courses—https://www.physalia-courses.org/). Over the same time frame, deep learning has become an ever larger part of our scientific work: from the application of neural network models to whole-genome predictions [1] for livestock and crop improvement, as well as in human and veterinary medicine (e.g. the genetic risk of disease), to research projects based on the development of computer vision models in crop phenomics (e.g. the *Polyploidbreeding* project, https://polyploidbreeding.ibba.cnr.it/) or in food science [2].

Caught between a small amount of (positive) pressure from the community and the opportunity to organise the material and expertise accumulated over the years, we cut the Gordian knot[1] and decided to write this book. Besides, everyone knows that we are living in the so-called A.I. era, the years in which Artificial Intelligence will either eventually rule the world and enslave us humans (as the most pessimistic fear), or finally lead humanity to the "promised land" of huge gains with little effort (as the most optimistic envision, with a touch of naïveté). Jokes aside, A.I. is all around us: not a single day goes by without countless media headlines on the perils and wonders of artificial intelligence. Most, if not all, of the spectacular A.I. applications we experience in our everyday lives (like the smartphone that

[1] Legend has it that, upon entering Asia, Alexander the Great was presented with a complex knot that tied an oxcart to a post: whoever could untie the knot would be destined to rule the entire continent. Alexander cut through it with his sword and proceeded to conquer Asia.

F. Biscarini, N. Nazzicari, *Deep Learning for Life Sciences*, Decoding Evolution 1,
https://doi.org/10.1007/978-3-031-96852-5_1

recognises and understands your voice) are based on neural networks and deep learning. What a fantastic moment to write a book on deep learning!

In the next box ("Deep learning applied to life science: a few examples"), we mention a few examples of applications of deep learning to science and technology. This is just the tip of the iceberg, however, and no such list could ever be complete. It is now clear that deep learning is both pervasive and here to stay for a long time. For a recent review of DL applications in medicine and healthcare see Chiranjib Chakraborty et al. [3]. Therefore, we think that students, scientists and professionals of the twenty-first century can not ignore this technology, but must come to grips with it and get at least a basic understanding of neural networks. This is the first step in moving away from the awe and panic when facing something new and mighty, and turning it into something manageable and useful. And that is precisely the objective of this book: to show you that neural networks are not black boxes, and explain you why. We'll start from a simple example and then build progressively more complex deep learning models, always describing with great care all the details of the structure and functioning of neural networks. At the end of the book, deep learning will have been completely demystified, and you will be able to understand the news and the literature, and to apply neural networks to your own work and research projects.

Deep Learning Applied to Life Science: A Few Examples

Lung cancer detection: given the importance of early diagnosis, and the global impact of the disease, there is strong interest in fast and reliable diagnostic tools. The development of automatic tools benefited from the availability of public databases of training data like LIDC [4], LUNA16 [5] and JSRT [6]. Approaches vary. The "standard" 2D and 3D *convolutional neural networks* train new medium-sized networks from scratch, that are thus specialized in lung cancer detection from the very beginning. A different approach is found with Natural Language Processing (adapted to the medical domain) and vision transformers, which start from very (very) big networks already trained on generic datasets (think of computer vision, chatGPT, very open ended problems). Those networks are then adapted to lung cancer detection using a technique called *fine tuning*. Whatever the approach, progress is rapid and performances are constantly improving [7–9].

Drug discovery is the problem of finding new suitable drugs, or repurposing already available ones, to treat a target disease. The problem is complex, and many factors need to be taken into account (drug-target interactions, side effects, drug synthesizability and so forth). Drug discovery is, overall, a success story for deep learning. This becomes apparent when examining scientific literature, and particularly in reviews: in 2018 deep learning was defined as a promising technique among many other alternatives [10]. In 2023 all groundbreaking discoveries in the field came from deep learning [11].

(continued)

Livestock Precision Farming: the general public may still retain a bucolic image of agriculture but, just like all other economic activities, this too is profoundly impacted by technology. One interesting application of deep learning to livestock production is the use of sound data for the early detection of respiratory problems in pigs [12]: in a nutshell, microphones are strategically placed on a farm so as to continuously record the background sounds of the pigs breathing. A neural network model analyses these data and establishes the sound pattern of healthy pigs: deviations from this pattern alert the farmers of a potential outbreak of a respiratory disease. This technology has already been successfully applied in a commercial product (SoundTalks®, https://www.soundtalks.com/).

1.2 Structure of the Book

The general structure of this book is outlined below:

Chapter 1 is the introduction, and you are almost done reading it. Good job!

Chapter 2: What is statistical learning? covers the basics. We explain what changed (and what didn't) when moving from a classical statistical perspective to deep and—more generally—machine learning. We also introduce important dichotomies like "regression vs. classification" and "supervised vs. unsupervised".

Chapter 3: Measuring the error: validation of learning models introduces one of the core aspects of all statistical/machine/deep learning problems: how to measure model performance. This chapter is not just a list of different error metrics (even if those are present). We go into the motivations behind the choice of a metric, and how to adapt a selected metric to the specific needs of your problem.

Chapter 4: Neural networks and deep learning decoded covers the core concepts of neural networks. We introduce the various types of layers, the meaning of optimization and the dangers of overfitting. Dense (a.k.a. fully connected) and convolutional neural networks (CNN) are described here.

Chapter 5: Recurrent Neural Networks (RNN) is about a specific type of network where information is ordered in a sequence, an order that matters. RNNs are particularly useful for analyzing streams of data, being it natural language, sound or a meteorological time series.

Chapter 6: Making it work is about tips and tricks. We present and discuss the most common advice on how to prepare your data and tweak your network to maximise the results.

Chapter 7: Deep learning research and applications: a few interesting stories is a very special chapter collecting interviews with international experts that use deep learning in their cutting-edge research work. We asked them to share advice on how to better use this technology, as well as their opinion on current and future trends in their field. A mock interview to large language models is also included and summarised in this chapter.

Appendix A—Python notebooks collects the code that accompanies many practical examples throughout the whole book.

Appendix B—Natural Language Models Interviews contains the full-length answers to some general questions on deep learning that we asked the current top-ranked Large Language Models available for free.

1.3 A Wee Bit of Notation

In this section, we'll try to bring some order to the chaotic realm of the technical notation used throughout the book. We will clarify the basic terms, jargon and notations that you—the reader—will encounter. Let's start with the title: you'll see that we'll use the terms **deep learning**, **neural networks** and **deep neural network models** (**DNN**) somewhat interchangeably. A few nuances notwithstanding, these terms can for all practical purposes be regarded as synonyms: this is how we'll use them in this book. Additional jargon that will be used in an interchangeable way includes: (i) **example**, or individual record, observation, sample; (ii) **response** variable, a.k.a. **outcome**, **dependent** or **target** variable, or **label** (for classification problems); (iii) similarly, the x predictors can be called **independent** or **explanatory** variables, **factors**, **covariates** or **features** (the latter is the preferred deep learning jargon); (iv) **validation** and **test** set in cross-validation; (v) **loss** and **cost** function in model optimization.

As for the mathematical notation, in most cases this is self-explanatory or well-documented in each specific section, as for instance in Part I. A list of the most common notations used in this book is given below (no need to read it all now: you can refer to it later on, as you progress with the reading of the book). However, a few things ought to be pointed out. First, you'll see that in Part I we'll refer to model coefficients as β's, as is standard practice in the statistical field. However later on we'll switch to the w notation for *weights*, which is standard in deep learning. These w will still, at times, be called model coefficients or parameters. The letter u will usually indicate a **unit** in the neural network, and l a **layer** of the network. Activation values (for record i) are denoted as a_i: these are the output of the *intermediate* or *hidden* layers of the network, and are sometimes denoted as h_i.

Most Common Mathematical Notations and Abbreviations

- y_i: the observed response variable for record i
- $\hat{y}_i$: the predicted response variable for record i

(continued)

- z: (linear) combination of inputs
- $g(z)$: activation function of z
- $h_i = a_i$: output of intermediate (hidden) layers of the neural network for record i (activation values)
- $\mathbf{W}$: matrix of model weights w
- $\mathbf{W}_{\mathbf{m,u}}^{[\mathbf{l}]}$: matrix weights w for m features, u units, at layer l
- $\mathrm{L}(\mathbf{y}, \hat{\mathbf{y}})$: loss function
- $x^{<t>}$: element at position t in a sequence

- predictive machine: synonym for predictive model in machine learning
- classifier: predictive model for classification problems
- NN: neural network
- DNN: deep neural network
- DL: deep learning
- ML: machine learning
- CNN: convolutional neural network
- RNN: recurrent neural network

Part I

From Statistics to Statistical Learning

2 What Is Statistical Learning?

There's a dialogue that happens at least a couple of times in each edition of our course on deep learning. It goes like this:

– [Student] My problem is <*something very complicated* >, I measured a lot of stuff and I want to know the role of <*a very specific variable* >. Can deep learning help me to do that?
– [Us, the teachers, with a tight smile] No.

This is of course a simplification, and in fact it's also half a lie. But before digging into the details of deep learning and its potential, we need to take a closer look at the student's question. What does it imply?

In our experience, we see that there's a lot of confusion around the exact meaning of the term "learning", and the question "what the heck is statistical learning?" is often left lingering in the background.

In traditional statistics courses, students are taught that statistics can be used to investigate the natural phenomena they are studying. As an example, medical research provides one of the most classical scientific questions: is drug A more effective than drug B for treating a specific illness? But every subfield of life science offers similar questions. In plant and animal breeding, we may ask if plants of variety A are better suited than plants of variety B to adapt and thrive in a specific environment. Think about your field of work and you can easily formulate a variation on this theme.

We humans learn how nature works. It's a life-long experience, and we have an intuitive understanding that there are two main approaches to learning. The first one can be summarized as "monkey see, monkey do". You start by observing how other people do something. Then, you do it yourself, fail, try again, and in the end you are able to obtain some results. The second approach to learning passes through the intellectual understanding of a phenomenon. It usually involves a teacher, a book, or a website that explains the rules of the game to you. In less abstract terms: some

F. Biscarini, N. Nazzicari, *Deep Learning for Life Sciences*, Decoding Evolution 1,
https://doi.org/10.1007/978-3-031-96852-5_2

people learn to play guitar by ear, trying to reproduce a tune they have in mind. Other people study music theory.

This has a surprising parallel in data science, where a lot of the debate is centered around trying to define the exact borders of two key concepts: *inference* and *prediction*. Inference focuses on understanding the rules, ideally as a step toward the formulation of an underlying theory. Focusing on prediction means that we just want to build a tool capable of telling us something about the future. Inference is about understanding, prediction is about making it work.

Back to our students, as a result of their basic statistics training, the urgence of inquiring about the effects of specific variables is often deeply rooted. We learnt that drug A is more effective than drug B for treating the disease, but what is the effect of age on the drug's efficacy? What about the patient's sex? Ethnicity? Average blood pressure? Hair color? Movie preference? Which variables should we use—and which ones should we throw out—to model drug efficacy?

These questions come straight from classical statistics, where they have a legitimate place. However, in statistical learning they are ill-posed. Often the model learns independently which variables to use and how, and it does it in a way that becomes inaccessible to humans. The model has become a so-called *black box*. Even if it is able to e.g. take into account that the zodiac signs have no influence on the efficacy of an experimental drug, this information is kept hidden away in its belly.

This, for sure, sounds like a limitation, but it turns out that it's not as bad as you may think. The interpretability of single variables progressively loses relevance as the number of variables increases, along with the number of their many possible transformations and combinations. And in modern science there are usually *a lot* of variables to be considered.

In this chapter we aim to demystify the confusion around statistical learning, attempt to give our view on it and then investigate what moving from traditional statistics to statistical learning (and akin) implies.

You Can Do Statistical Learning in Your Head!

Let's say that the first edition of this book sells 10 copies—an unexpected success! The second edition sells 20 copies, the third edition 40 copies and the fourth edition 80 copies. (This may remind the learned reader of brahmin Sissa, who invented the game of chess and requested of the King that he'd be rewarded with one grain of wheat on the first square of the chess-board, an amount that is then doubled on every subsequent square—eventually outstripping the King's resources (https://en.wikipedia.org/wiki/Wheat_and_chessboard_problem). With this book, will we outrun Springer's printing capacity?) Question: how many copies of the fifth edition are likely to be sold? Think about it for a moment, then keep reading.

(continued)

The solution is obviously 160 copies. What may be less obvious is the underlying mathematical model:

$$c = 10 * 2^{(e-1)}$$

where c is the number of copies sold, and e is the edition number. You probably just picked up the obvious "always doubling" pattern, without formalizing an explicit mathematical model in your head, but you were able to give the correct answer anyway. For a moment, you became black boxed.

2.1 Analysis of Variance, Linear Models and Learning

One of the datasets used for teaching at The University of Sheffield shows that the birthweight of children from mothers who are smokers tends to be lower compared to that of children of non-smokers.[1] The distribution of birthweights, in kilograms, in the two groups is illustrated in the boxplots of Fig. 2.1. By calculating the well-known sums of squares from basic statistics [13], the analysis of variance (ANOVA) tells us that the difference in children's birthweights between smoking and non-smoking mothers has a p-value below the "magical" 0.05 threshold and we conclude that it is statistically significant (Table 2.1 on the following page): the smoking habit of the mother does influence the birthweight of her children.[2] As a matter of fact, the traditional ANOVA sum-of-squares-based analysis is equivalent to a linear model of the following form (Eq. 2.1):

$$\text{bw}_{ij} = \beta_0 + \beta_1 \cdot \text{smoker}_j + e_{ij} \tag{2.1}$$

where bw_{ij} is the i_{th} child's birthweight (kg) from mother j; smoker_j is an indicator variable that takes up the value of 1 if the mother smokes, and 0 otherwise; e_{ij} is the model residual; β_0 and β_1 are, respectively, the intercept (overall mean) and the coefficient for the "smoker" effect.

[1] This dataset can be downloaded from https://www.sheffield.ac.uk/mash/statistics/datasets (From U. Sheffield's website: The birthweights of the babies whose mothers smoked have been adjusted slightly to exaggerate the differences so that students can see them more clearly in a plot).

[2] Long and yet unresolved debates have populated -and will undoubtedly continue to do so- discussions in the scientific community over the 0.05 significance threshold for experimental results. Rather than feuding over specific significance thresholds, we personally believe that moving to empirical resampling or Bayesian probabilistic approaches is by far more meaningful. This said, p-values will not be particularly relevant in this book.

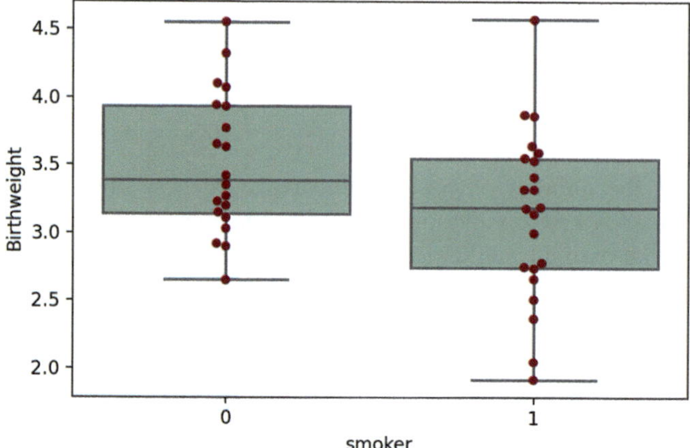

Fig. 2.1 Birthweight of children—in kilograms—delivered by smokers (1) and non-smokers (0)

Table 2.1 ANOVA table

Term	SS	d.f.	F	Pr(>F)
Intercept	246.33	1	731.18	2.5e−27
Smoker	1.48	1	4.38	0.042
Residual	13.48	40		

By solving model 2.1,[3] we can obtain the solutions for $\beta_0 = 3.51$ and $\beta_1 = -0.375$: these can be interpreted as the average birthweight of children of non-smoking mothers, and the average difference in children's birthweights between smoking and non-smoking mothers. We can thus conclude that on average we expect children from smoking mothers to weigh 375 g less than children from non-smoking moms.

This is a very simple model, and it makes intuitive sense that there are more factors involved in determining the birthweight of a newborn other than whether or not the mother smokes. We can thus go on and add a second variable to our linear model, for instance the weight of the mother before pregnancy (pre-pregnancy weight, `mppwt`):

$$\text{bw}_{ijk} = \beta_0 + \beta_1 \cdot \text{smoker}_j + \beta_2 \cdot \text{mppwt}_k + e_{ijk} \tag{2.2}$$

We solve model 2.2 and see that this second variable is also significant (p-value = 0.006) and has an estimated coefficient of 0.034, which can be interpreted as

[3] To solve this kind of problem you can use standard statistical approaches to parameter estimation like the *least squares* method. If you are not familiar with this don't worry. We'll go into the details as soon as they are really needed. For now, just consider that the proposed solution (i.e. the proposed values for β_0 and β_1) is the one that best fits the data or, in other words, implies the lowest amount of error.

follows: for every additional kilogram of the mother's weight (before pregnancy) the child's birthweight increases on average by 34 g (*ceteris paribus*). We can keep adding variables, for example we can ask ourselves if the age of the mother (`mage`) has an effect on the child's birthweight. We add a third term to our model, solve it and find out that the p-value for `mage` is 0.65: it is not statistically significant and we decide not to include the age of the mother in our model. But what if the relationship between the age of the mother and the child's birthweight is not linear? We can try to incorporate polynomial terms (i.e. the square and the cube of the mother's age) into our model (*keep in mind that this is still a linear model!*). We also simplify the notation with respect to the subscripts, using only i to indicate individual data records:

$$\begin{aligned} \text{bw}_i = \beta_0 + \beta_1 \cdot \text{smoker}_i + \beta_2 \cdot \text{mppwt}_i + \beta_3 \cdot \texttt{mage}_i \\ + \beta_4 \cdot \texttt{mage}_i^2 + \beta_5 \cdot \texttt{mage}_i^3 + e_i \end{aligned} \tag{2.3}$$

Solving model 2.3, we find that the polynomial terms (transformed mothers' ages) are now somewhat statistically significant, with p-values 0.071, 0.051 and 0.047: should we include them in the model? And what about their coefficients? These are -1.74 for mage, 0.065 for mage^2+ and -0.00079 for mage^3: how should we interpret them? Loosely speaking, we can say that, on average, we lose 1.74 kg of birthweight per year of the mother's age,[4] then we add 65 g per squared year of mother's age, and ultimately we lose 0.8 g per cubed year of the mother's age (actually, it is a little more complicated than this: we could argue that the quadratic term can also be interpreted as the change of slope for each additional unit of x, sort of a second-order derivative; and the sign of the coefficient for the quadratic term would indicate concavity—or convexity).

Clearly, the interpretation of our model becomes a bit contrived. And we could go on and on adding variables to our model (there are 14 variables besides the child's birthweight in the `birthweight` dataset). We could transform variables as we please (polynomial terms, roots, logarithms or any other functions of the variables), and we could combine them into interaction terms (e.g. the interaction between the age of the mother and that of the father). You can easily see that it will soon become a titanic task to decide which variables and functions of variables (and combinations thereof) to keep in or remove from the model, and the interpretation of the model in terms of the specific role and meaning of each variable will be virtually impossible.

As we remarked at the beginning of Chap. 2 (What is statistical learning?), the questions of variable selection (which variables should I include in the model?) and model interpretability (what does each variable mean and what role does it play?) quickly and progressively lose meaning in the context of statistical learning, at least

[4] With the polynomial model the intercept changes to 16 kg, which accommodates for a loss of almost two kilograms per extra mother's year.

in their traditional form (and more so in deep learning). We find ourselves moving in the field of predictions, rather than in that of inference. This calls for a change in perspective and in the way we approach the evaluation of a statistical learning model. The focus is mainly on the performance of the model: does it or does it not work? It seems that we are back to the *black box* approach, but don't despair: we will be able to take a peek inside the box!

2.2 Function Approximation: At the Heart of Learning

Let's play a simple game. Each day you offer us either an apple or a banana. If we like the offer we give you a gold coin. If we dont't like the offer you get nothing. We assure you that there's a strict logic governing our daily taste in fruits, but we refuse to explain any further.

At the beginning, your understanding of the game is very poor. You pick the type of fruit randomly. Sometimes you get a coin, sometimes not. This goes on until you start taking notes and notice that bananas have a slightly higher chance of receiving a coin. In fact, 55% of the times it's a banana day, and only 45% of the times it is an apple day.

Reassured by this breakthrough, but still moved further by your greed, you start taking notes on other facts. You notice that the bias toward banana days is true in general, with the exception of Sundays. On Sundays most of the times it's an apple day.

You keep taking notes, and start noticing finer patterns. When we meet you while sitting on the sofa there's a slight advantage for apples, while when you find us standing at the window there's a strong advantage for bananas.

After a long time, you have a collection of rules and subrules. The very basic strategy of always offering a banana[5] gives you an average of 0.55 gold coins per day. But if you apply all the rules in your notebook you reach an astounding 0.81 average gold coins per day. That's a clear improvement!

What happened?

It happened that you learned more and more rules of the game, building a model (your notes) that describes with a growing level of precision a hidden function (our secret rules). The approximation is not perfect—on average you still miss 19% of the available coins—but the improvement is there.

Can you do better? What is the maximum precision achievable? Can you get 100% accuracy?[6]

[5] Introducing a bit of nomenclature, in this example bananas and apples are *classes*, and, in particular, banana is the *majority class*, since it is observed more often. Here we are talking about a *balanced problem*, since the differences between the two classes (55% vs. 45%) is not very big. Conversely, an *unbalanced problem* would have a very different class ratio (e.g. 90% probability of having a banana day) and would require some special attention.

[6] In the field of statistical learning, and particularly when talking about classification problems, the word *accuracy* indicates a specific way of measuring performance. It is computed by dividing the

It depends. Maybe we have been unfair. Maybe one of the secret rules is "each third Monday of the month we flip a coin. If we get heads it's a banana day. Tails, and it's an apple day". That's a valid rule, well defined and unambiguous. It's also playing dirty. First, it would be very difficult to *pick up its signal*. In other words it's hard to see the small perturbation in patterns caused by this rule, especially if you are not analyzing the data in the framework of "first Monday of the month, second Monday of the month...".

Second, it limits the maximum theoretical accuracy. There is one day of the month when you cannot do better than 50–50. Do the math and you'll discover that, in a thirty-days month, the maximum achievable accuracy becomes 98.33%. Still high, but not 100%.

Many real world problems share similarities with the banana-apple guessing game:

- given age, biological sex, body-mass index, blood pressure, heartbeats per minute under stress, and other tens of observed parameters, will a patient have a heart attack in the next year?
- given the genetic content read at 52,041 known SNP locations by a commercially available soybean beadChip, what would be the yield per acre of a specific soy variety?
- given the ordered temporal sequence of COVID-19[7] cases, can we forecast the evolution of the epidemic? How many cases next week? And next month?
- given the basic list of amino acids composing a protein, will its zinc finger structure bind with Poly (ADP-ribose) polymerases?
- ... and many more

As with the banana-apple guessing game, in the real world we collect data and are interested in predicting the next round. To get a sense of what's going on, we measure the performance of our predictions (we will soon introduce different metrics other than accuracy). We build models, which hopefully improve over time, and which are going to be our best approximation of the *underlying, secret, and ultimately unknowable* set of rules that command the game.

Neural networks—which are the main topic of this book—are very good, we'd say excellent, at function approximation (approximating the 'secret rules' or 'hidden patterns' behind the problem). First of all they are very generic, meaning that almost any kind of problem can be fed to a neural network regardless of the nature of the input (continuous values, alphanumeric strings, categories, images, sounds, 3D

---

number of times the model guessed the correct answer (a.k.a. the *Correct Predictions* count) by the total number of predictions.

[7] The regrettably famous "Coronavirus disease 2019".

scans, long texts, videos etc.)[8] or of the expected output (yes/no answers, numeric predictions, probabilities, summary of a text, new synthetic images, etc.).

Second, neural networks—especially deep neural networks—are forever improving approximators, meaning that if you can feed them more data their performances usually improve. This is no small feat. Many other statistical models often encounter some sort of saturation phenomenon where new data don't really affect the accuracy very much.[9] Neural networks are always hungry for data, and a diligent scientist can take advantage of this feature.

2.3 From Artificial Intelligence to Deep Learning

Even in its simplest form, a thermostat implements a crude but effective logic: if the temperature falls below a certain desired value the heating is turned on. If the temperature rises above that same threshold the heating is turned off. Can we say that the thermostat is a form of artificial intelligence?

Strictly speaking: yes, there's little room for arguing. No debate around the *artificial* part, since no thermostats are found in nature. Regarding the *intelligence* part, it is very limited, true, but it's there nonetheless, since it can be argued that a thermostat is clearly more intelligent than any completely-passive device. In fact, if you add the option to set the temperature via an app, it will be marketed as being a "smart thermostat". Smart and intelligent really sound like synonyms.

Unfortunately, the definition of Artificial Intelligence that descends from the literal meaning of the words ends up incorporating everything more sophisticated than a screwdriver. It also contrasts with another very popular meaning, the one usually evoked when people talk about androids and machines that think like-humans. The confusion is so great that the need arose for a more meaningful acronym: AGI, Artificial General Intelligence.[10]

Inside the AGI borders you find artificial sentience and consciousness, plus the capability to adapt and the ability to solve very different kinds of problems. And, of the utmost importance for what we cover in this book, the capacity for autonomous

---

[8] We'll later see, though, that all input data need somehow to be converted to a numeric representation to be read by the neural network: data representation is an important area of deep learning.

[9] See the discussion led by Andrew Ng on how scale drives deep learning: https://www.youtube.com/watch?v=LcfLo7YP8O4.

[10] The AGI acronym was presented and popularized by Ben Goertzel and Cassio Pennachin in their 2007 book "Artificial general intelligence" [14]. Goertzel admits: "*I didn't love it tremendously, but I fairly soon came to the conclusion it was better than any of the alternative suggestions.*" He also rejects the primacy for being the first to come up with term, and cites a semi-obscure 1997 article from Mark Gubrud on "Nanotechnology and International Security" as the first instance of somebody talking about Artificial General Intelligence. For the interested reader, the original blog post from Ben Goertzel is archived at https://web.archive.org/web/20181228083048/http://goertzel.org/who-coined-the-term-agi/.

learning: artificially intelligent systems must be able, when posed with a problem, to find solutions on their own, and improve on the initial solution by progressively learning to perform a given task better. The observant reader may have noticed that, while thermostats are commonly available, we don't have any such sentient machines (yet?), and thus their exact features remain uncertain at best.

The talk about artificial intelligence (general or not) focuses on what a device or a piece of software can or cannot do. This can lead to confusion, since the perimeters of some tasks are ill defined and there is ample room for debate. A more concrete set of categories focuses instead on the underlying mathematical models, and has some direct link to the main topic of this book. From the more general to the more specific, we have:

- **statistical learning**: is the broadest category, traditionally focusing on statistical models still understandable by humans. In this sense, it is often said that statistical learning offers *inference*, meaning that the model allows for an improved understanding of the physical phenomenon.[11]
  A very simple example of statistical learning is computing the average of a set of values. It tells us something about the set, and offers a rough prediction of where the next element of the set will be (i.e. somewhere close to the mean).
- **machine learning** is a subgroup of statistical learning in the sense that it still uses mathematical models to extract and rationalize information, but the models have crossed the limit of human understandability. They can be used for predictions, but offer limited inference.
  An example of machine learning is given by Support Vector Machines [15], which transform the data into a different representation that usually loses all intuitive links to the original problem. The assumption here is that the mathematics works better in this transformed state and intractable problems can become approachable.
- **Neural Networks** are a type of mathematical models, and we'll talk a lot about them in this book. In the very simplest form (Single Layer Perceptron), they still make some intuitive sense, but as soon as they grow (Multi Layer Perceptron and up) it becomes almost impossible to follow their internal logic exactly. This makes Neural Networks a fully-fledged member of the Machine Learning family.
- **Deep Neural Networks** are a subcategory of Neural Networks. They are different from "standard" neural networks from a topological point of view—trivially, they contain more layers. More importantly, Deep Neural Networks can leverage both large and unstructured (loosely formatted) datasets.[12] This is a

[11] Statistical and machine learning can often be considered to be quasi-synonyms, depending on whether the focus is more on statistics or on programming.

[12] *Big data* refers to large, fast and complex datasets characterized by the 3 V's (Doug Laney, 2001): Volume (large quantity), Velocity (fast growth), Variety (data type and form). The latter also includes the so-called *unstructured data*, which are data that don't come from a specific data model or are not organised in a pre-defined manner (e.g. text data, sound data, etc.). More on this later on.

clear advantage compared to traditional, "shallow" Neural Networks and is part of the big success of Deep Neural Networks in the recent years.

2.4 Supervised vs Unsupervised Learning Problems

For our numerical illustrations, so far, we have used: (i) the number of copies of the book sold each year; (ii) the birthweight of children as a function of the smoking habit of their mothers (plus additional characteristics of the mother); (iii) whether or not a coin was obtained depending on the choice of fruit. These are all examples of *supervised learning problems* (the learning process is supervised by the availability of the answers). For each measurement of the predictor variable x_i (e.g. year number, mother's smoking habit, choice of fruit) we have the associated response measurement y_i (e.g. number of copies of the book sold, child birthweight, reception of a coin—or not). We can then learn a function that maps x values to y responses with the aim of predicting new observations, thereby "approximating" the -unknown- underlying true function $x \mapsto y$: see Sect. 2.2 (Function approximation: at the heart of learning) and Fig. 2.2. The majority of this book will cover this type of supervised learning problems, at which deep learning excels.

However, there is a large class of generally more challenging problems for which we have a set of measurements x_i with no associated responses y_i. These are called *unsupervised learning problems*: there is no predetermined answer available to guide our (or, better said, the machine's) learning process! In unsupervised learning the aim is to find relationships between the variables or between the observations, while moving blindly around the data. On the "pro" side, data collection for unsupervised learning problems is generally simpler, since there's no need to obtain labels (or response values) for a large training set. Broadly speaking, unsupervised learning problems include: (i) clustering (grouping together records with similar characteristics); (ii) pattern recognition (identifying patterns in the data); (iii) anomaly detection (identifying datapoints that don't belong to the dataset); (iv)

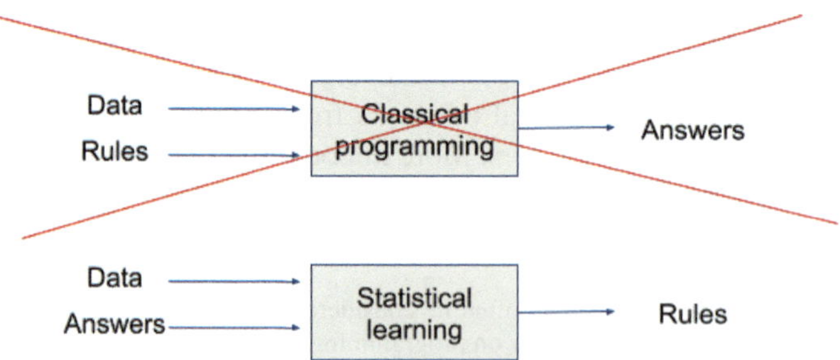

Fig. 2.2 Classical programming vs the learning paradigm. In classical programming data and rules are used to give answers; in statistical/machine/deep learning, data and answers are used to learn the rules (and then try them out on new data). This holds for supervised learning problems, for which answers are available. What if we don't have answers?

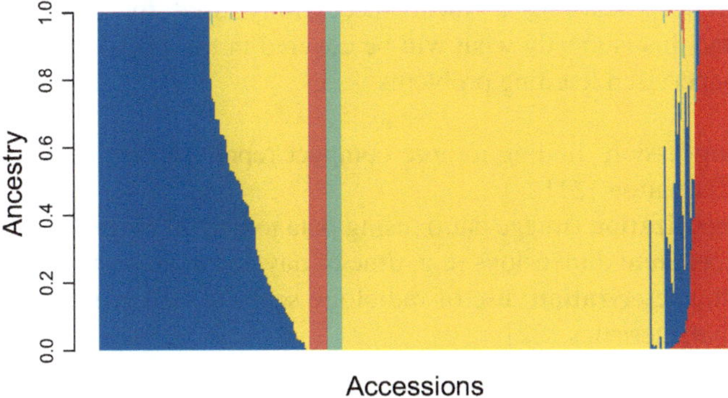

Fig. 2.3 Stacked barplot for the ancestry of 391 rice (*Oryza sativa*) varieties with K = 5 (K is the number of clusters). Ancestry groupings were based on 166K SNP (single nucleotide polymorphisms) genetic data

dimensionality reduction (e.g. compressing most of the information into a smaller number of variables/dimensions).

Clustering is the most common type of unsupervised learning problems: for instance, one might want to group customers based on their shopping data so as to customize promotions and discounts (i.e. market segmentation), or analyse unstructured text data to categorize texts into a number of coherent topics (subjects) or to identify the topics covered by any given text (a set of techniques known as *topic modelling* [16]). An historical example lies at the foundation of epidemiology: in the 1850s, during a cholera outbreak in London, the physician **John Snow** plotted the location of cholera-related deaths on a map of the city, thereby discovering that cases tended to cluster around contaminated sources of water. This finding allowed scientists to understand the cause of cholera cases and the possible solutions [17]. In current biology, one example is the analysis of gene expression data from cancer patients to find coherent subgroups of patients (subtypes of the disease) or of genes (thus leading to a better understanding of the underlying mechanisms of the disease). Another common problem is the clustering of individuals based on their genetic ancestry, using genomic data: Fig. 2.3 shows the ancestry of rice (*Oryza sativa*) varieties based on single nucleotide polymorphism (SNP) data [18]. In the following box ('The perilous waters of unsupervised clustering'), we provide a detailed description of a clustering problem, together with a *Jupyter* notebook with solved and explained *Python* code.

Beyond clustering, the popular and widely used **Principal Component Analysis** (PCA) is a technique aimed at transforming the data to find a reduced number of dimensions (named the *principal components*) that capture most of the information contained in the original data (by the way, similar approaches are used by data compression algorithms [19]). In an example known as the cocktail party problem, the words spoken by a person need to be separated from the background noise (signal recognition) [20]. In fraud detection, unusual transactions need to be identified from a large number of credit-card use data (anomaly detection).

Although deep learning is most successfully used for supervised learning problems (and this is mostly what will be covered in this book), it can also be used to solve unsupervised learning problems:

- **data compression**: finding a more compact representation of the data without losing information [21]
- **photo organization** (image data): using data to gather images in a gallery based on some coherent dimensions (e.g. time of day, location, semantic meaning)
- **tumor characterization**: use of radiology scans to find groups of tumors with similar characteristics [22]

Additionally, unsupervised learning is at the centre of a very active field of research that aims to train models with partial data. In many practical problems, it is unrealistic to obtain large labeled training datasets, since labeling is typically a human activity that turns out to be expensive and time-consuming. You can think, for instance, of labelling images based on their content (e.g. given a photo, which object/animal is represented in it?), or of diagnosing patients for a binary classification problem (given some medical data, is the patient sick or not?). In these examples, it is potentially easy to obtain thousands (or even millions) of photos or medical records (e.g. radiological scans, laboratory tests, etc.), but it would be difficult and take a long time to assign the correct labels to each record. In such circumstances, it would be very helpful to be able to train the model based on a few labeled records, but at the same time leveraging the many available unlabeled records: this is called "semi-supervised learning" [23], and is an example of a promising application of deep learning to unsupervised problems. The neural network will make use of both the few labeled and the many unlabeled records to find patterns in the data and correctly associate them with the corresponding labels (whether observed or not). After training on very large datasets, such models are expected to perform outstandingly well in practical applications (unseen new data).[13]

The Perilous Waters of Unsupervised Clustering

You are given the map in Fig. 2.4 and a simple task: you need to color all the islands belonging to the same archipelago using the same color. You are also given three crayons—a red one, a blue one and a green one. The task seems trivial, with little room for ambiguity.

(continued)

[13] There's more fish in the ocean of learning: besides supervised, unsupervised and semi-supervised learning, we find reinforcement learning, active learning, adversarial learning, transfer learning etc. Reviews can be found in [24] and [25].

Fig. 2.4 An unsupervised classification dataset. Map generated using Azgaar's map generator, https://azgaar.github.io/Fantasy-Map-Generator/

You are then asked to repeat the task, but this time you receive only two crayons. Islands will be either red or blue. Even if at face value the problem could be considered simpler—with fewer options to be selected—you are now actually facing difficult choices. Either you split one of the three archipelagos in half or you put two of them together. In any case, there is no clear solution, and each decision is suboptimal.

The tasks keep coming, and the number of crayons received changes each time. You get four, five, six and more colors, and must use all of them.

The optimal number of clusters[14] doesn't depend on the type of problem—coloring groups of islands—but it is bound to change with the specific dataset. A different map may contain four, five or just one group of islands.

Often clustering algorithms require the number of clusters as an input (usually indicated by the letter K). In other words, when you apply an algorithm to a new dataset, you need to decide how many crayons will be used. There are three common sources for K:

- **field knowledge**: if you have some prior understanding of the problem
- **business decision**: if the number of clusters is somewhat imposed by the project
- **elbow method**: a common heuristic for estimating the best fitting value for K

(continued)

One of the simplest but also most effective unsupervised clustering algorithms is called *K-means clustering*. To observe it in action on the above map, together with an implementation of the elbow method, see Sect. A.1.

[14]A *cluster* is a subgroup of elements in which in-group variability is on average lower than out-group variability. All elements of a cluster are more similar to each other than to members of other clusters. Each cluster has a "centroid", i.e. the ideal average member of the subgroup. In the map example the centroid of each cluster would be a point on the map representing the average position of the islands for that subgroup. Centroids allow us to understand what elements are more central to the cluster and what elements are more peripheral.

2.5 Regression and Classification Problems: What's the Difference?

Classification problems are about dividing your samples into groups. Regression problems are about assigning a number to each sample. These are both types of supervised problems, and the thing that we want to model and ultimately predict is the response variable y. This is the output of our statistical machine, and it is in general the most important aspect of our problem.

A few examples:

- given a set of audio recordings of kiwi birds calls, separate the male calls from the female calls. This is a classification problem.
- given biological sex, age, and height of the parents, predict the height of the child. This is a regression problem.
- given the genomic profile of a set of white lupin (*Lupinus albus*) lines, predict which ones will be subject to anthracnosis (a fungal disease) and which ones will resist. This could be either a classification problem (if the response variable is coded as affected/not affected) or a regression problem (if the response variable contains information about how much the plant is affected by the fungus).

Regression and classification problems call for different mathematical functions, and in fact some classification algorithms (e.g. decision trees[15]) have nothing in common with traditional regression models (e.g. standard linear regression models). We are however interested in bridging the gap between the two worlds. In particular,

[15] A decision tree is an intuitive algorithm that takes one example and assigns it to one or the other "branch" based on the value of a given measured variable. This is done over many variables (successive branches of the tree) with all examples ultimately assigned to one of many possible final nodes ("leaves"). Decision trees are typically used for classification problems, but can just as effectively be applied to regression problems [26].

we are interested in using the vast literature available on regression problems and adapting it to classification problems.

A simple way to do that is through another mathematical function that takes as input the continuous response variable outputted by a regression model and then "discretizes" it. The simplest case happens when we want to use a regression model to distinguish between two classes (e.g. class A and class B). It goes like this:

1. take your input variables X and feed them to a regression model $y = f(X)$. The output will be, in the most general case, a number $y \in [+\infty, -\infty]$.
   - In other words, y will be a single real number, unbounded.
2. feed y to another function $z = g(y)$, carefully built so that $z \in [0, +1]$
   - In other words, z is a positive number between zero and one
3. interpret z as the probability of the sample belonging to the first or second class:
   - the closer z is to zero, the more probable that the sample belongs to class A
   - the closer z is to one, the more probable that the sample belongs to class B
   - a result of $z = 0.5$ means complete uncertainty
4. The classification model is the combination of the two functions: $z = g(f(X))$
5. If a probability is still not enough and you actually need a class, you can round z to the closest integer.

This is the general recipe for transforming a regression model in a binary classification model. The same approach can be easily extended to more than two classes. We'll talk about this topic in more detail in Sect. 4.4 (Logistic regression from a neural network perspective: binomial and multiclass classification problems) and Sect. 4.5 (From logistic/softmax regression to neural networks), where we'll discuss logistic regression, binomial and multinomial classification and the transition to neural network models.

2.6 A Small Taxonomy of Learning

We can summarize what we have discussed so far in oversimplified Table 2.2.

Table 2.2 Hypercompact taxonomy of the types of learning problems

		Are labelled examples available? Yes (supervised learning)	 No (unsupervised learning)
Response variable	Discrete	Classification	Clustering
	Continuous	Regression	???

The parallel between classification and clustering is straightforward: in both cases the problem consists in assigning each sample to a group. When doing classification the list of groups is known a priori and a number of samples have already been assigned. When doing clustering the groups are unknown—maybe even the very number of groups needs to be decided—and no samples have been assigned to any group yet.

When talking about continuous response variables, the parallel is less immediate. Regression problems are very well known, but it's not clear what an "unsupervised regression" would look like. Still, there are interesting unsupervised applications that are worth mentioning:

- **Dimensionality reduction**: when, for each sample, many variables have been measured, it may be necessary to prune the data and keep only useful variables (feature selection) or to create aggregate indexes that summarize in a single number the information of more than one variable
  - **Autoencoders** are the deep neural network take of the general *Dimensionality reduction* task, where a network is trained to build a very compact representation of the original data
- **Pattern recognition** is the automated recognition of patterns and regularities in data. The output is usually a list of (the most common) patterns and optionally their position in the original dataset.
  - **Association** is a type of pattern recognition that highlights what types of data point are most often found together. An example is noticing that, in the English language, the letters "t" and "h" are often found together. Another example is that some items, like beer and diapers, are often sold together in supermarkets.[16]
- **Anomaly detection**: based on the available X features, for each sample the probability of belonging or not to the specific data distribution at hand can be calculated. This applies to discrete as well as continuous response variables (even when these are not directly available, as in unsupervised problems)

With this last step we have almost concluded Part I and are now ready to dive into proper Neural Networks, but only after we have done a brief excursus into the realm of metrics used to measure model performance in the next section.

[16] The association between beers and diapers has an almost legendary status in the field of *Market Basket Analysis*. Mark Madsen traces the story of this discovery in his presentation "Beer, Diapers, and Correlation: A Tale of Ambiguity", illustrating many of the often overlooked limitations of this kind of results. For a survey of traditional MBA techniques see [27], for an investigation of the difficulties linked to the data dimensionality see [28].

Measuring the Error: Validation of Learning Models 3

In Chap. 2 (What is statistical learning?), we saw that a key feature of learning models (be it statistical, machine or deep learning) is their ability to figure out (learn) on their own which variables to use and how. For the model to be able to learn, we need to be able to measure its performance: how well—or badly—is the model doing? Learning is an incremental process, and at each iteration the model will do a little better or a little worse: if it did better, it will continue on to the next step; if it did worse, it will go back to the previous step and try something different in the hope of doing a little better.

But how can we measure if the model did better or worse, and by how much? For this we'll need to have **error metrics**, i.e. a way to measure the performance of the learning model that we are developing. In supervised learning problems, we already have the answers alongside the measured variables x_i, so it is straightforward to come up with sensible error metrics: we let the model learn on its own, then we compare its results with the (known) answers and see how close or far from them it is. Depending on the type of problem (regression, classification, forecasting) the specific metrics will be different, but the concept remains unaltered: we look at the difference between model results (the predictions) and the known answers (response variables, labels). In unsupervised learning problems things obviously get more difficult: this is a treacherous ground on which to move, since the true answers are not known. Still, not all hope is lost, and ways to measure the performance of unsupervised learning models can be defined.

In this chapter, the main focus will be on describing the "ins and outs" and the "pros and cons" of the most important error metrics used in statistical/machine/deep learning methods for **supervised problems**. A section will be dedicated to the more arduous task of measuring performance in unsupervised learning problems. Additionally, we will describe a general framework for correctly and unbiasedly measuring the performance of a learning model: **cross validation**.

F. Biscarini, N. Nazzicari, *Deep Learning for Life Sciences*, Decoding Evolution 1,
https://doi.org/10.1007/978-3-031-96852-5_3

3.1 Error Metrics

We will now see some of the most common metrics used to measure the performances of a predictive machine. Remember that we can either face classification problems (e.g. what's the race of this dog?) or regression problems (e.g. how tall will a baby be once he/she becomes an adult?). Given that the nature of the desired predictions are different, the performances are measured in different ways.

Error Metrics for Classification Problems

We now introduce a bit of notation, that will be used all throughout this section. We have:

- X : the input variable(s), i.e. the data that you want to use to make a prediction, be it a CT scan of a patient, the genetic profile of a plant, a recording of a bird's mating call or some economic indicator
- $y \in \{a, b, c...\}$: the output variable, i.e. what you want to predict. Since we are considering classification problems here, y will associate the sample with one of the available classes (a, b, c ...). There is no limit to the number of classes, but it must be finite. However, in practical terms, a classification problem with millions of classes is rarely seen.
- $\hat{y} = f(X)$: the actual prediction, i.e. the output of the predictive machine $f()$ once it's fed with some new data. Please keep in mind that the only way to measure the performance of a predictive machine is to "predict" data that you already have, so that you can verify the answers.

The above notation is relative to a single sample (e.g. a single picture of a dog whose breed you want to determine; a single microscope image of a cell that needs to be classified as either normal or cancerous; and so forth). However, only very poor statistics can be obtained using a single data point. Therefore all symbols should acquire a subscript "i", e.g. X_i, y_i, $\hat{y}_i$, with $i \in \{1, 2, ..., n\}$, where n is the number of samples that we'll use for measuring the performances of the predictive machine.[1]

Error Rate (and Its Limitations)

The simplest metric for a classification problem is called the **error rate** and is computed as the number of prediction errors over the total number of predictions. Formally:

$$\text{ERROR RATE} := \frac{1}{n} \sum_{i=1}^{n} cnt(y_i, \hat{y}_i)$$

[1] This group of samples is called a *validation set* or, in some specific cases, a *test set*. It is separated from the *training set*, which contains all the data points used to tune the predictive machine. For more details see Sect. 3.2 (Cross-validation and resampling methods).

with the counter function $cnt()$ defined as:

$$cnt(y_i, \hat{y}_i) = \begin{cases} 1 \text{ if } \hat{y}_i \neq y_i \\ 0 \text{ if } \hat{y}_i = y_i \end{cases}$$

This metric is very intuitive.[2] We can measure it and then report that a classifier (i.e. a predictive model for classification) has an error rate of 0.25: it is immediately clear that three times out of four its predictions are correct, and a quarter of the times it makes a mistake.

By definition, the error rate is bound in the [0, 1] interval, meaning that it ranges from zero (no classification errors) to one (nothing was correctly classified).[3]

However, the error rate has a strong limitation in problems in which a specific class is rare compared to the other(s). Consider a system that, given the DNA profile of a patient, will detect the presence of a rare disease. Let's say that 99% of the general population doesn't have the disease. To understand the limits of the error rate we are going to compare two predictive machines: an honest one and a dumb one.

The honest predictor is trying to detect the disease. In fact, three quarters of the patients that actually have the disease will be detected. Unfortunately no system is perfect, and out of every ten healthy patients two are erroneously classified as carrying the disease.[4]

Doing the math:

$$\underset{\text{(honest classifier)}}{\text{ERROR RATE}} = 0.01 * 25\% + 0.99 * 20\% = 0.2005$$

Let's now consider a classifier that doesn't even try to predict anything. Since most people don't have the disease, this classifier always labels the patient as healthy. This results in:

$$\underset{\text{(dumb classifier)}}{\text{ERROR RATE}} = 0.01 * 100\% + 0.99 * 0\% = 0.01$$

If we just look at the error rate we should conclude that the second classifier is doing a better job. This is clearly an undesirable result, since that classifier is not really doing anything and all the patients carrying the disease get misdiagnosed. The

[2] You can sometimes find this alternative notation for the error rate: $Err = \frac{1}{n}\sum_{i=1}^{n} I(y_i \neq \hat{y}_i)$, where $I()$ is an indicator function that returns 1 if the prediction is wrong ($y_i \neq \hat{y}_i$), 0 if it is correct ($y_i = \hat{y}_i$).

[3] The complement of the error rate is called *Accuracy*, and it is simply computed as 1—Error Rate. Its intuitive meaning is the opposite: it is desirable to have high accuracy (and thus a low error rate).

[4] These are known as *False Positives*, while the opposite type of error (a sick patient labelled as healthy) is a *False Negative*. This nomenclature is valid only with binary classification, and loses meaning when three or more classes are present.

same limitations are present if the problem has three or more classes with a single dominant class that contains the majority of the samples. In other words, the error rate is a bad metric when it is important to take into account rare cases. It is thus necessary to explore different metrics that are more robust to class unbalancedness. We'll start by looking at the **confusion matrix** (see the box "Don't be confused by the Confusion Matrix" below).

Don't Be Confused by the Confusion Matrix

A *Confusion Matrix* is a synthetic way to represent the performance of a classifier. It is easily extendable to multiclass problems, and can give interesting insights on the types of errors that the model is making. At its core, a Confusion Matrix is a square matrix with as many rows and columns as there are classes in the problem. On the main diagonal the number of samples correctly classified is reported. On the off-diagonal cells we find the number of errors, i.e. samples that were of one class and got erroneously classified as belonging to another class.

Take a look at the example of a Confusion Matrix reported in Table 3.1, where the problem is to classify pictures of three types of fruits: apples, bananas and tomatoes. Looking at the table it is immediately clear that the classifier often mistakes apples for tomatoes, and vice versa, while very few errors are made when classifying bananas.

There are many variations in the way a confusion matrix is presented. It is possible, for example, to normalize the table by dividing each cell by the total number of samples. Sometimes totals by row and by columns are presented, thus reporting the classes and the attributions tally. And in case of many classes color coding can be useful, thus transforming the table into a heat map. Many statistical libraries have standard functions to compile and render a heatmap. More details on this can be found in Sect. A.2 (Notebook—confusion matrix)

Table 3.1 The confusion matrix for a fruit classification problem. The first row tells us that ten Apples were correctly classified as such, one was classified as a Banana, and four as Tomatoes. The other rows follow the same logic. By column, the first tells us that of the 16 predicted apples, 10 were indeed apples, while 6 were actually tomatoes. And so on

		$\hat{y}$ (predicted labels)		
		Apple	Banana	Tomato
y (true labels)	Apple	10	1	4
	Banana	0	8	1
	Tomato	6	0	11

Crossentropy

By far the most common performance metric used for classification problems in the field of deep learning is crossentropy. The reasons for this function's popularity are outside the scope of this chapter and are related to its use as a loss function in model optimization (see Sect. 4.7 (Optimization)).[5]

To explain how crossentropy is computed we need to introduce the actual output of a predictive machine. Recalling the example of the previous section on the error rate, we start with a model that needs to predict if a patient is healthy (H) or sick (S). We test our model on the following set of four patients:

	y (true label)
Patient 1	H
Patient 2	H
Patient 3	S
Patient 4	H

Predictive models usually don't output actual labels. Instead, the output of a classifier will look like the following:

	y	$\hat{y}$	
	(true label)	Prob(H)	Prob(S)
Patient 1	H	0.7	0.3
Patient 2	H	0.6	0.4
Patient 3	S	0.1	0.9
Patient 4	H	0.2	0.8

That is, the model outputs the probability that the sample belongs to each available class. This is also true in the case of multiclass problems, as long as the sum of the probabilities for each sample adds up to one. If at the end it is necessary to actually assign a label to a sample, it will be usually enough to choose the class with the highest probability.

In the table above, we see that for the first row the model is 70% sure that "Patient 1" is healthy. The second patient is also considered healthy, but with a lower probability. The model is very sure that the third patient is sick, but unfortunately on the last patient it makes a mistake.

[5] For the impatient reader: the metric of choice needs to make the total loss function differentiable (which the error rate/accuracy is not). Crossentropy is not the only available metric, but among the viable ones presents a good combination of ease of computation and stability of results.

The general target for machine learning is to maximize the probabilities so that the model is as sure as possible about the correct labels. A high confidence on the correct answer is preferable to a low-confidence correct answer, and this latter in turn is preferable to any wrong answers. The worst case happens when a model is very sure about a wrong answer.

In this framework, the crossentropy is defined as the negative of the sum over all samples of the natural logarithms of the probabilities of the true class. In the example above the crossentropy would result in:

$$\underset{(example)}{CROSSENTROPY} = -1 * [ln(0.7) + ln(0.6) + ln(0.9) + ln(0.2)]$$
$$\simeq 2.582298996 \tag{3.1}$$

Note that each probability is bound between zero and one, and the logarithm of such a quantity is a negative number. The change of sign (the -1 at the beginning of the formula) thus ensures that crossentropy as a whole has a positive value (and can thus be minimised).

To get an intuitive sense of why crossentropy is a good metric, remember that if the model is completely sure about the correct prediction we obtain $ln(1) = 0$ and if it's completely sure about the wrong prediction we obtain $ln(0) \rightarrow -\infty$. In other words, the crossentropy will be lower (i.e. closer to zero) the more confident the model is about the correct answers. However, the more errors the model makes, the more the crossentropy will grow, with no upper bound. This is an important difference compared, e.g., to error rate, which is bound between zero and one. More errors can always make the crossentropy grow more. In this context, a predictive machine will thus try to minimize its total crossentropy.

When Labels Are Not Certain

It may happen that even the y (True Labels) are probabilistic, thus reflecting a state of uncertainty over the available data. Or maybe an intrinsic feature of the problem makes the labels look like probabilities. For example Fig. 2.3 shows the case of a set of rice accessions, each one with its ancestry split among five subpopulations. There is no uncertainty there: a mixed ancestry is just that, mixed.

In these cases both y (True Labels) and $\hat{y}$ (Predicted Labels) consist, for each sample, of a set of values between zero and one and adding to one. We can then use the following generalized crossentropy definition:

$$\underset{(generalized)}{CROSSENTROPY} = \sum_{i}^{\text{samples}} \sum_{j}^{\text{classes}} -1 * y_{ij} * ln(\hat{y}_{ij})$$

(continued)

This formulation is similar to the one illustrated for non-probabilistic labels, and follows the same logic. As a matter of fact, this generalized crossentropy equation can be reduced to the standard equation (3.1) if we don't have uncertainty about the true labels. In this case y will be entirely composed of ones (for the correct class) and zeros (for the wrong class), thus simplifying the equation.

ROC Curves and AUC (Area Under the Curve)

We go back to the confusion matrix, now focusing on the simple (and by far the most common) case of a binary classification problem: we therefore have two classes, "healthy" and "sick" subjects. The objective of the classifier is obviously to use some variables X to predict whether any given subject has or does not have the diseases. The results are in Table 3.2. From this confusion matrix we can of course calculate the error rate, as described in the Section "Error rate (and its limitations)": there are 10 errors in 100 predictions, hence $\text{Error rate} = \frac{10}{100} = 0.10 = 10\%$. We already discussed the limitations of the basic error rate, and for this reason four additional error metrics can be derived from the confusion matrix.

We can look at the table from the perspective of the actual values, that is by looking at the total number of healthy subjects (the negative cases) and at the total number of sick subjects (the positive cases): in this way we get the **false positive rate** (**FPR**) and the **false negative rate** (**FNR**), which express the error rates in the two classes separately (together with their "accuracy" counterparts: the true positive and true negative rates: $TPR = 1 - FNR$, and $TNR = 1 - FPR$).

We can then change perspective and look at the confusion matrix by the perspective of predictions (column-wise), i.e. by considering the total numbers of positive and negative predictions. In this way, we obtain the **false omission rate** (**FOR**, i.e. the percentage of false negative predictions), and the **false discovery rate** (**FDR**, i.e. the percentage of false positive predictions), which express the error rates in the two types of predictions (together with their counterparts: the positive and negative predictive values: $PPV = 1 - FDR$, and $NPV = 1 - FOR$).

Table 3.2 Confusion matrix for a binary classification problem: TN: true negatives; FP: false positives; FN = false negatives; TP = true positives. The related major error rates are also reported: FPR (False positive rate) $= \frac{3}{78} = 0.038 = 3.8\%$; FNR (False negative rate) $= \frac{7}{22} = 0.318 = 31.8\%$; FOR (False omission rate) $= \frac{7}{82} = 0.085 = 8.5\%$; FDR (False discovery rate) $= \frac{3}{18} = 0.167 = 16.7\%$

		(Predictions)		
		Healthy	Sick	
(Reference)	Healthy	TN = 75	FP = 3	FPR
	Sick	FN = 7	TP = 15	FNR
		FOR	FDR	

When the data are unbalanced, like in Table 3.2 (22 positive cases, 78 negative cases: ratio close to 1/4), we already know that the accuracy is a misleading metric. From these data, the accuracy is 90%, but this breaks down to TPR = $\frac{15}{22} = 68.2\%$ and TNR = $\frac{75}{78} = 96.2\%$. We see that there is a large difference between the accuracy in the two classes (there are more negative examples in the data, therefore the tendency is for TNR to be larger than TPR), and we learn that it is always a good idea to look at both TPR and TNR when evaluating a model's performance. However, when we want to optimise a model—e.g. via fine-tuning of a complexity hyperparameter through cross-validation—we need one single metric or criterion to use as target. It would therefore be very convenient if we could combine TPR and TNR in a single metric: the **ROC curve** does exactly this! ROC stands for "receiver operating characteristic":[6] based on probability values ($P(y = 1|X)$: binary classification), all possible classification thresholds in [0,1] are used to produce the confusion matrix and calculate TPR and TNR (remember: the most common threshold is 0.5, i.e. the example is classified as a positive case if $P(y = 1|X) > 0.5$, as a negative case otherwise). Therefore, we have many TPR and TNR values corresponding to all possible thresholds, and the idea of the ROC curve is to plot them in relation to one another. Actually, the ROC curve plots TPR on the y-axis, and 1-TNR (FPR) on the x-axis (Fig. 3.1). Intuitively, we see that points on the upper-left corner indicate good classification performance, with both high TPR and low FPR (hence high TNR). The ROC curve can be summarised by one single value, the **AUC** (area under the curve): if for all classification thresholds we perfectly classify all positive and negative cases, the AUC is equal to 1. Points along the diagonal (white line in Fig. 3.1) mark the area where AUC = 0.5, i.e. the classifier has no predictive ability and yields random classifications ("chance accuracy": in other words, our classifier is no better than the flipping of a coin—unbiased).

The ROC AUC is therefore an excellent metric for classification problems (binary classification: however, extensions to multiclass classification problems do exist [30–32]), that summarize the accuracy in both classes and makes a trade-off between TPR and TNR (thereby avoiding models that look deceptively good by predicting well only in the majority class, i.e. "accuracy blunders"). Yet, the AUC is not devoid of drawbacks, and under some extreme circumstances can be a misleading measure of the model performance [33]. As we have seen, the AUC only takes into account two of the four basic ratios from the confusion matrix: TPR and FPR = 1-TNR (the accuracy measured on the true labels). However, when data are strongly imbalanced this can be suboptimal: if, for instance, the number of positive examples (one class) greatly exceeds the number of negative examples (the other class), a sizeable change in the number of false negatives (positive examples wrongly classified as negative

[6] The term "receiver operating characteristic" was developed during World War II for radar signal detection to distinguish between true signals and noise. "Receiver" is the electronic receiver used in radar systems to detect signals; "operating characteristic" refers to the graphical representation of the performance of a binary classifier system as its discrimination threshold is varied [29].

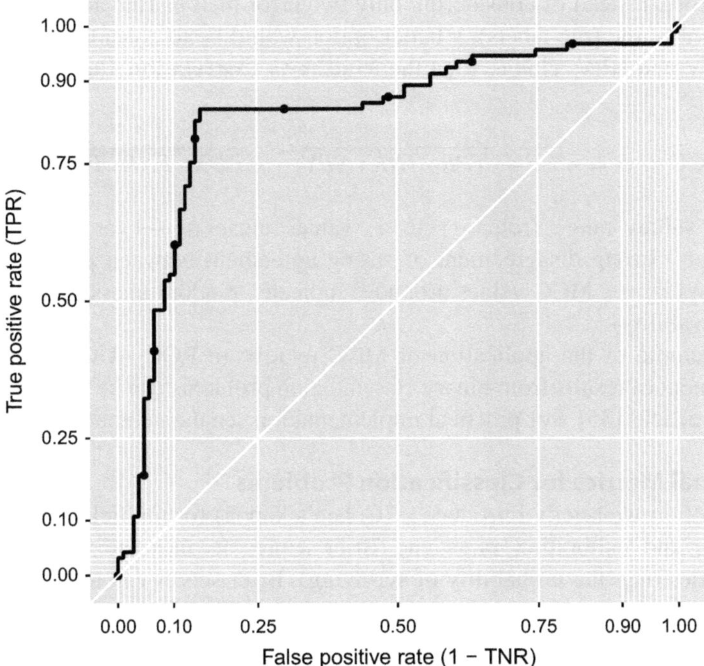

Fig. 3.1 Example of a ROC curve for a binary classification problem. The diagonal white line corresponds to chance accuracy, i.e. 50% probability of belonging to either class (equivalent to the flipping of a coin, total absence of predictive ability of the classifier). This ROC curve is for illustration only and is based on mock data; it doesn't correspond to the results (also made-up) in Table 3.2 (for which we have no probability values)

cases) can lead to a small change in the false negative rate used in the calculations of AUC (the denominator is given by the sum of the many true positives and the relatively few false negatives). On the other hand, under the same extreme conditions, the negative predictive value (NPV) would capture the effect on the model's performance of the large number of positive examples compared to negative examples, by comparing false negatives with true negatives rather than true positives (in other words, by restricting the calculations within negative predictions only). For example, if we have 1020 'cases' (positive examples) and make twelve errors (FN = 12), the TPR would be $\frac{1008}{1020} = 0.988 = 98.8\%$. If we change something in the classifier, and the number of this type of errors increases by 50% (new FN = 18), the TPR would decrease to 98.2%, an almost imperceptible change. And the FPR would remain unchanged (only the number of FN has increased), thereby producing almost no change in ROC-AUC. However, with the same results and, say, 20 true negatives (remember, this is the highly unbalanced minority class), the NPV would go from 62.5% to 52.6%: this sharp decrease would hardly go unnoticed!

Therefore, instead of considering only two error metrics (the accuracy from the perspective of the true labels), a better option would be to consider all four: TPR, TNR, PPV and NPV. This is what the **Matthews Correlation Coefficient** (MCC) does [34]:

$$\text{MCC} = \phi = \sqrt{PPV \cdot TPR \cdot TNR \cdot NPV} - \sqrt{FDR \cdot FNR \cdot FPR \cdot FOR}$$

MCC values range from -1 to 1: values close to -1 or to $+1$ indicate, respectively, strong disagreement or strong agreement between predicted classes and actual classes. MCC values around 0 indicate "random guessing" (absence of predictive ability).[7]

An example of the application of MCC relative to ROC-AUC for the correct measurement of results from binary classification problems can be found in a recent scientific article [35]. For practical implementation see the notebook in Sect. A.3

Additional Metrics for Classification Problems

In this section we briefly introduce κ (Cohen's kappa) for multiclass classification problems, and calibration plots and Brier scores to measure the accuracy of probabilities (e.g. the probability of surviving). Brier scores can also be expressed relative to a naive benchmark (e.g. prevalence of a disease; average survival by age class etc.).

For multiclass classification problems it is not straightforward to use the same metrics that we saw for binary problems (TPR, TNR, PPV, NPV). In this case we can instead use **Cohen's kappa** (κ) to measure model performance [36]. To calculate Cohen's κ we need two ingredients: (i) the **accuracy** of the model ($1-$ Error rate); (ii) the **chance accuracy**, which is obtained from the relative frequencies of observed and predicted classes over all examples. Then Cohen's κ is calculated as:

$$\kappa = \frac{\text{accuracy} - \text{chance\_accuracy}}{1 - \text{chance\_accuracy}}$$

As you see, the numerator of the equation calculates the difference between accuracy and chance accuracy:

- if accuracy $= 1$, we have perfect predictive performance (the confusion matrix is diagonal), regardless of chance accuracy (numerator and denominator are the same, and $\kappa = 1$)
- if accuracy $>$ chance accuracy, then $\kappa > 0$ (positive), and it means that the model is doing better than chance

[7] MCC can also be calculated directly from the confusion matrix: $\phi = \frac{(TP \cdot TN - FP \cdot FN)}{\sqrt{(TP+FP)\cdot(TP+FN)\cdot(TN+FP)\cdot(TN+FN)}}$.

- if accuracy = chance accuracy, then $\kappa = 0$ (the numerator is zero), and the model is on average as good as pure chance
- if accuracy < chance accuracy, then $\kappa < 0$ (negative), and the accuracy of the model is lower than what it would be by chance

When we work with probabilities, besides looking at the predicted outcomes (e.g. positive and negative predictions against observations summarised in a confusion matrix) it would be good to be able to assess the "accuracy" of the probabilities estimated by a model. This is especially important in cases where you calculate the probability of events like the survival of surgical patients at n years after the intervention, or rainfall in a specific place and time. A relatively straightforward way of assessing the quality (accuracy) of model predictions is to compare them with the observed rate (proportion) of predicted events: if our model predicts that 80% of young patients will survive 10 years after heart surgery, we should see that approximately 80% of young patients have survived 10 years after surgery, for our model to be good. Similarly, if the model predicts 60% survival for older patients, we should observe about 60% survival in older subjects (again, if the model has good predictive ability). In other words, a model is **well-calibrated** if its predictions match the observed distribution of event rates.

Model calibration can be summarised in a **calibration plot**, as shown in Fig. 3.2: on the x-axis we have the model probabilities (binned by 0.2 steps—5 bins); on the y-axis we have the event rates for the examples included in each bin. The colored dots represent the event rate that corresponds to the probability bins, and the bars around the dots are the 95% confidence intervals. The dashed-gray line indicates perfect calibration: at each probability level corresponds exactly the same event rate (e.g. 20% predicted survival rate $\rightarrow$ 20% observed survival rate). We can see that the model in Fig. 3.2 is reasonably well calibrated, if a little underbiased: when model probabilities are low, the event rate is low, when probabilities are high, the event rate is also high (if you want to get more information on the topic, there is obviously much more to model calibration [37]).

Calibration plots are a great visual tool for assessing the accuracy of model probabilities. However, just as was the case with ROC curves, it would be handy to get a single number to express the overall performance of a probability model. As you can guess, such a single number does exist: it's the **Brier score**, originally devised by the American statistician Glenn Brier for weather forecasts [38]. Basically, the logic behind the Brier score is similar to the mean squared error (see the Section "Error metrics for regression problems" for details), but applied to probabilities rather than continuous values. For each example in the dataset, we take the difference between the model probability ($P(y = 1|x)$) and the class label (0 or 1) and we square it. We then sum all squared differences together and normalise by the number of examples. Mathematically:

$$\text{Brier score} = \frac{1}{N} \cdot \sum_{i=1}^{N} \left(p(x)_i - y_i)^2 \right)$$

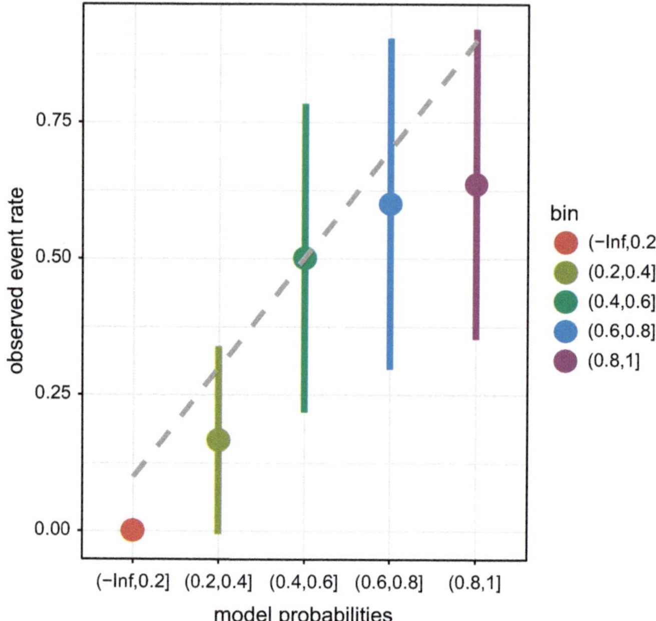

Fig. 3.2 Example of a calibration plot for a binary classification problem. The diagonal gray dashed line corresponds to "perfect calibration", i.e. the estimated model probability exactly matches the rate of the event. The dots represent the event rate for any given probability level; the bars represent the 95% confidence interval around the event rate: $\pm 1.96 \cdot \sqrt{\frac{\hat{p}(1-\hat{p})}{N}}$, where $\hat{p}$ is the event rate and N is the number of observations used to calculate the event rate

For the data in Fig. 3.2, the **Brier score** is 0.171. Being an error rate, the lower the Brier score, the better (a value of zero would indicate perfect probabilities: always one for cases, always zero for controls). This said, the Brier score suffers from a somewhat difficult interpretation: it is not immediately easy to tell whether a specific Brier score value is high or low, or—more importantly—if we can conclude that our probability model has performed well. This is why it is a good practice to compare model probabilities with a benchmark. Let's say that our event (again, Fig. 3.2) is a disease: we could naively use the disease prevalence as a benchmark probability tool, i.e. if the prevalence is 30% we can assign $P(y = 1|x) = 0.3$ to all subjects in our study. We can now calculate the Brier score for our benchmark probability, and we get 0.216: this is higher than 0.171, which tells us that using the prevalence to make predictions results in a worse outcome than our probability model (as expected). Now, we can express our Brier score relative to the benchmark: $100 \cdot \frac{(0.171-0.216)}{0.216} = -20.4\%$. This means that our model improved predictions by 20.4% over the benchmark!

More details on the calculation of Cohen's kappa, calibration plots and Brier scores can be found in the accompanying notebook that you find in Sect. A.4.

Table 3.3 Common metrics for measuring the error of regression models, when the target variable is continuous

Metric	Formula
Mean Squared Error (MSE)	$\frac{1}{n}\sum_i^n (y_i - \hat{y}_i)^2$
Root Mean Square Error (RMSE)	$\sqrt{\frac{1}{n}\sum_i^n (y_i - \hat{y}_i)^2}$
Mean Absolute Error (MAE)	$\frac{1}{n}\sum_i^n \lvert y_i - \hat{y}_i \rvert$
Mean Absolute Relative Error (MARE)	$\frac{1}{n}\sum_i^n \left\lvert \frac{y_i - \hat{y}_i}{y_i} \right\rvert$

Error Metrics for Regression Problems

When facing regression problems the y values (the True Answers: a.k.a. response, target, outcome or dependent variables) are numbers, and it is thus not possible to keep using concepts, like the *number of errors*, that have been used with classification. That's because no predictive machine will be able to predict the exact values of the analyzed samples up to an arbitrary precision.

Fortunately, the problem of measuring the performance of a regression model has been very well studied and there are several possible solutions (and it is actually somewhat easier compared to classification models).

Metrics Based on Error

All methods in this group use, at their core, the difference between true and predicted answers, i.e. they compute $y - \hat{y}$. Table 3.3 summarizes the four most common metrics (here n is the number of samples in the test set).

All metrics are based on the difference between true and predicted values. Since a prediction can either under- or over-estimate y, the error can be a positive or a negative quantity. All metrics compensate for this fact, either by squaring or taking the absolute value of the difference. Note that the final value is always positive. Thus, a predictive machine will try to minimize every one of these measures.

MSE gives the average squared error, which is difficult to interpret (but has excellent properties as a loss function for model optimization: more on this—a lot more—later); RMSE (and MAE as well) brings back the average error on the scale of the y (response variable), which is more readily interpretable. MARE lies in $[0, 1]$ and gives an error proportion (or percentage if you multiply by 100).

Metrics Based on Similarity

For some applications it may be of interest to measure the prediction performance using a general description of similarity between true and predicted values. The most common example of such similarity metrics is the **Pearson's correlation coefficient** ($r_{y,\hat{y}}$), given by the following formula:[8]

[8] The $n - 1$ in the equation stems from the concept of *degrees of freedom*. In very simple terms, when you calculate deviations from the mean (both at the numerator and denominator: covariance and standard deviations), all numbers in the series are theoretically free to take up any value except the last one, which is constrained by the known value of the mean.

$$r_{y,\hat{y}} = \frac{1}{n-1} \sum_{i}^{n} \left(\frac{y_i - Y}{s_y} \right) \left(\frac{\hat{y}_i - \hat{Y}}{s_{\hat{y}}} \right)$$

where:

- n is the size of the validation set
- Y, $\hat{Y}$ are the sample means of true and predicted values, respectively
- s_y, $s_{\hat{y}}$ are the sample standard deviations of true and predicted values, respectively

Pearson correlation is bound between the $+1$ (perfect correlation) and -1 (perfect anticorrelation). Values around zero indicate no correlation whatsoever.

It must be noted that Pearson correlation is a metric of **linear correlation**: although there may be not many reasons to think that the relationship between continuous y values and the corresponding predictions $\hat{y}$ is not linear, it is useful to know that measures of non-linear correlation do exist (e.g. distance correlations, mutual information, maximal information content [39]).

Besides the correlations between observed and predicted y values, another common approach is to calculate the correlations between their ranks, i.e. the order in which they are sorted from highest to lowest values: it may well be that the numerical values are not very close (predictions tend to differ from observations), yet their relative order is maintained, i.e. the top observations remain at the top and the bottom observations remain at the bottom. The **Spearman correlation coefficient** calculates the rank correlation [40]: it's still in the range $[-1, +1]$, with $+1$ indicating perfectly equal rank, -1 indicating perfectly opposite rank, and 0 indicating random ranking of predicted values. Looking at the preservation of ranks rather than actual values can be valuable in many circumstances: for instance, in plant and animal breeding, when you try to predict the hidden genetic value for a desired phenotype (e.g. yield, or disease resistance), the actual number is not important, but conversely the order is of paramount importance: you want to be sure that the plant (or animal) with the highest genetic value is correctly predicted, so that it can be selected and used to breed the next generation with expected increased genetic merit and, hence, phenotypic performance. Interestingly, the Spearman correlation does not hinge on the assumption of linearity, and can also be applied to sets of variables that have a non-linear (yet monotonic) relationship.

Another valuable metric is the **Normalized Discounted Cumulative Gain** (NDCG), which is based on samples ranking [41]: this is similar to Spearman correlation, but instead of considering the whole rank it focuses on partial ranks, i.e. the top (or bottom) portion of the rank. In the field of information retrieval, it is often used to measure the effectiveness of web search engine algorithms or related applications.

This metric is suitable when it's important to maintain the position of the top samples, i.e. the samples with the highest values for y should have the highest values for $\hat{y}$. The importance given to each sample is scaled down (discounted) through the ranking, so that the last samples give little contribution to the metric. In other words, errors in ranking of the top samples have a much greater effect compared to errors in the low ranking samples. The same line of reasoning holds, of course, if we are instead interested in the bottom part of the rank. For the NDCG formula, together with working examples, please see Sect. A.4

3.2 Cross-validation and Resampling Methods

A predictive model is evaluated on the quality of its predictions. But any predictive model also needs to be tuned: we can think of a global "complexity" (hyper)parameter that can be geared towards simpler or more sophisticated models. This implies that the available experimental data can be used for either purpose, i.e. for training the model or for testing its performance. This separation of purposes needs to be clear, and is one of the most sacred principles in statistical learning. No piece of data used for training a model should *ever* be used to evaluate it. Violating this principle will lead to inflated accuracies and abnormally low error rates due to the model, in a sense, somehow "cheating", since it already knows the test before the exam.

More formally, for our predictive model we are interested in estimating the quantity P, which represents its performance as measured using any of the metrics we have seen in the previous sections (e.g. crossentropy, Mean Squared Error and so forth). The true value of P is unknowable, since it would require testing the model against all possible data sets of interest. We more modestly aim to build an estimator $\bar{P}$ that approximates P, and we want this approximation to be as good (low bias, low variance) as possible.

We will illustrate the two main strategies for managing sample data within the framework of training predictive models: crossvalidation and repeated resampling.

Cross-validation is technically a very broad term that refers to all situations when part of the data is used for training and another part is used for testing. For all practical purposes when people talk about cross-validation they usually refer to "K-folds cross-validation", a technique that goes like this:

1. decide K, with common choices being 5 and 10. Let's say we pick $K = 5$
2. randomly split the samples into K equally sized groups. If the sample size is not a multiple of K try to make the groups as even-sized as possible
3. you now have K groups, which we'll call *folds*, and which will be referred to as $\{f_1, f_2, \dots f_K\}$. If $K = 5$ they go from f_1 to f_5
4. use samples from folds $\{f_2, f_3, \dots f_K\}$ to train the model. If $K = 5$, this corresponds to using 80% of your data for training
5. use samples from fold f_1 for testing and obtain the measure $\bar{P}_1$ of the model performance. If $K = 5$ this corresponds to using 20% of your data for testing
6. cycle through the previous two steps, each time selecting a different fold for testing and using the remaining $K - 1$ folds for training. If $K = 5$ this means repeating the training four other times using in turn f_2, f_3, f_4 and, f_5 for testing.

7. as a result of the previous steps you have computed a set of K measures of the model performances, $\{\bar{P_1}, \bar{P_2}, \ldots \bar{P_K}\}$. Taking the average we obtain $\bar{P}$, the K-folds crossvalidation estimator for P

The choice of K influences the training process. As K grows, the fraction of data used for training increases, which pushes toward model stability and a more faithful representation of the world. However growing K shrinks the test set, thus forcing the evaluation of the various $\bar{P_1}, \bar{P_2}, \ldots$ to be computed on smaller datasets. In this scenario the importance of each single sample grows, to the point that the presence of any outlier or anomaly in the data can heavily influence the results. In more formal terms, larger values of K reduce model bias at the cost of increasing its variance, as illustrated in Fig. 3.3. An extreme case is when K=N, which leads to **leave-one-out cross-validation** (LOOCV): in LOOCV each example is used in turn as validation "set" (validation data-point) while N-1 examples are used for training. On one hand, this means running N models (which, for models more complex than multiple linear regression, is going to be computationally intensive); on the other hand, all N training sets of size N-1 will be very similar, and hence highly correlated. Therefore, the variance of predictions will be pushed to the extreme, severely hampering the process of training and fine-tuning the predictive model. To give an example: in binary classification problems when the left-out validation data point belongs to the minority class, prediction rules like majority vote would always give the wrong results.

The Risk of Doing It Wrong

As the number of variables increases, the risk of overfitting the data becomes higher, thereby increasing the chance of finding models that look good on the training data but that will have a low predictive ability on independent data. In simple terms, overfitting means *fitting the data too well*: if we keep adding variables and their combinations, we can easily develop a complex model that perfectly describes the data. Such a model would fit the data "like a glove", but would lose the possibility of being generalised and applied to other (new) data, thereby becoming virtually useless for the understanding and explanation of the natural phenomenon at hand and for any practical applications. Overfitting will be further discussed in later sections of this book.

(continued)

▶

Fig. 3.3 Effect of the number of folds K applied to a linear regression problem. The dataset, plotted as gray dots, is randomly split into training and validation sets. The colored lines represent the different models obtained via Least Squares (3 regression curves for K=3, 5 regression curves for K=5, 10 regression curves for K=10). As the value of K grows the model is trained on a larger set, thus resulting in lower bias (the training set is larger) and larger variance (the validation set is smaller). This is reflected in the performance metrics (measured on the validation set). Pearson's correlation is transparent to bias (being linear) and shrinks as K grows (it is affected by variance).

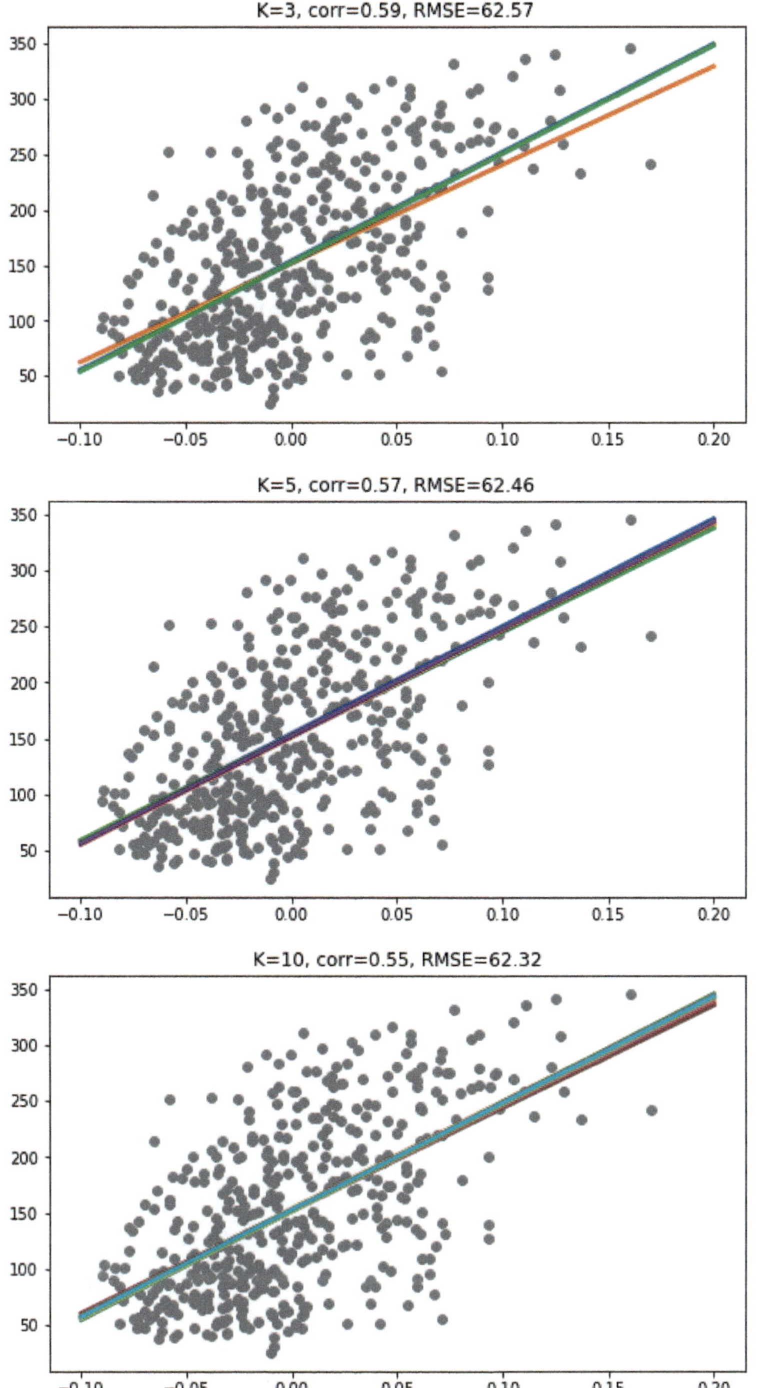

Fig. 3.3 (continued) The Root Mean Squared Error is more sensitive to bias than variance and improves (decreases) with larger values of K

We have seen that **k-fold cross-validation** is a powerful method to control overfitting. However, it is crucial that cross-validation is carried out in the "right way", otherwise it can lead to underestimating the prediction error. This brings us back to the "sancta sanctorum" of cross-validation: **the test (validation) set is untouchable!**

Consider a regression problem: 100 samples, 50,000 variables (e.g. 'omics data):

- Step 1: Find the 100 features with the strongest correlation with the response variable
- Step 2: Develop a predictive model with only these 100 selected features

We want to estimate the prediction error: **can we apply cross-validation in Step 2?** The short answer is: **NO!** In Step 1 the model has already used the response of the training data. The 100 selected variables have been "cherry picked" based on the data: this is already training, and the correlation with the response may be a result of the specific configuration of this dataset (a "quirk" in the data).

An illustration of the risk of doing the cross-validation wrong is provided by a controversy on genes reported to be predictive of survival in patients with follicular lymphoma [42].

You could test this yourself: simulate random data (no association between labels and variables: true error rate = 50%), then cherry-pick those variables most associated with the label and use these—and only these!—in a predictive model. It won't be difficult to build a model with error rate = 0 from cross-validation!

Take home message: when $p >> n$ (many more parameters p than observations n), it is very tempting to preselect variables based on their relation to the response variable, and then to apply cross-validation to the reduced dataset. **Resist the temptation!**

If you actually need to do variable selection, it should be done **inside** cross-validation. In other terms, split your data into training and test sets and select the important variables considering only the data in the training set. Then train the model. If you are doing K-fold cross-validation you'll end up doing the variable selection step K times, thus selecting different variables each time. You can then take a look at the most-often selected variables and declare those as the most promising ones. You can do whatever you want in the training data, but only after you have taken out the test/validation set!

Repeated sampling is a technique similar to K-folds cross-validation, but simpler. It starts with fixing a fraction $R \in]0, 1[$ (open real interval). The whole dataset is then split into training and validation sets following the selected value (e.g. if $R = 0.3$, the training set will contain 70% of the data, and the validation set 30%). The model is then trained and evaluated as usual. The random split is repeated a number of times, and several observations of the performance metric(s) are collected and eventually averaged.

Compared to cross-validation, this approach does not ensure that all samples will be part of the validation set (they become part of it only asymptotically as the number of repetitions grows towards infinity). Therefore, fewer data are used for training and testing, and this leads to more bias and more variance (hence, overall poorer performance). It does, however, have two advantages. Firstly, it allows for an arbitrary split between training and validation sets, while cross-validation is bound to the inverse of integer numbers, i.e. $K = 2$ implies a 50/50 split, $K = 3$ a 66/33 split, $K = 4$ a 75/25 split and so forth. Secondly, it's more easily implemented. In practical terms most deep learning software libraries offer some variation of *Repeated sampling* that works out of the box. Contrarily, to implement cross-validation the user is usually required to resort to some degree of data management.

The choice of R commands the size of the training and validation sets and as such has the same consequences on the bias/variance trade-off already discussed for cross-validation.

3.3 Unbalanced Datasets and Class Weights

In the Section "Error rate (and its limitations)", we discussed why the *error rate* is not a good metric for classification problems with unbalanced data sets (i.e. problems when one class has many more samples than the other). But even when using a more appropriate metric like *crossentropy*, unbalanced data sets need some special attention. Moreover, even when the problem is balanced, it is possible that the cost of errors is not. For example, you may wish a classifier that predicts the presence of a dangerous pollutant to err on the side of caution, since a false positive (i.e. an alarm risen when there's no danger) is not equivalent to a false negative (i.e. an alarm *not* risen when the danger is present).

In these cases it may help to provide the model with *class weights*. The exact details will be implementation-dependent, but most classifiers from modern libraries support them. The aim is self-explanatory: to tweak the model tuning so that it takes into account that not all errors are equal. The weights can either come from field expertise or be computed algorithmically. In the first case weights are produced after an evaluation of the application and the effects of having false positives *vs.* false negatives. In the second case some heuristic can be implemented to extract balanced weights. With unbalanced data sets a popular formula [43] is:

$$w_i = \frac{N}{C * n_i} \tag{3.2}$$

where w_i is the weight assigned to class i, N is the total number of samples, C is the total number of classes, and n_i is the number of samples in class i. The idea in this formula is that if a class contains the expected share of samples (e.g. half of the samples in a binary problem, a fifth of the samples in a five-classes problem) its corresponding w_i will be equal to one. If the class contains more than its share, it is considered over-represented and its weight will be smaller than one. In case of underrepresented classes the weight will be larger than one. Note that given that the weights are scaled with the inverse of the number of samples per class, the total sum of weights doesn't add up to one (or to any meaningful number).

3.4 Stratification: When Data Are Not Independent

In Sect. 3.2 we described how to correctly estimate the predictive ability of statistical models with cross-validation, and we also discussed the risks of doing it wrong (box "The risk of doing it wrong").

This is the basic scenario, when you get neat and tidy data from e.g. a single simple experiment. However, real-world datasets coming from large experiments or collated from multiple sources (e.g. different projects, field data, repeated sampling campaigns etc.) tend to be less homogeneous and to show complex internal dependence structures. In biological datasets, these dependencies usually fall within four categories: (i) spatial (geographically distributed data); (ii) temporal (e.g. time-series data); (iii) treatment (e.g. experimental groups); (iv) hierarchical (e.g. genetic ancestry/population structure) (Fig. 3.4).

Statistical models can be designed to account for such dependencies (e.g. autoregressive models, mixed models). However, in cross-validation these dependence

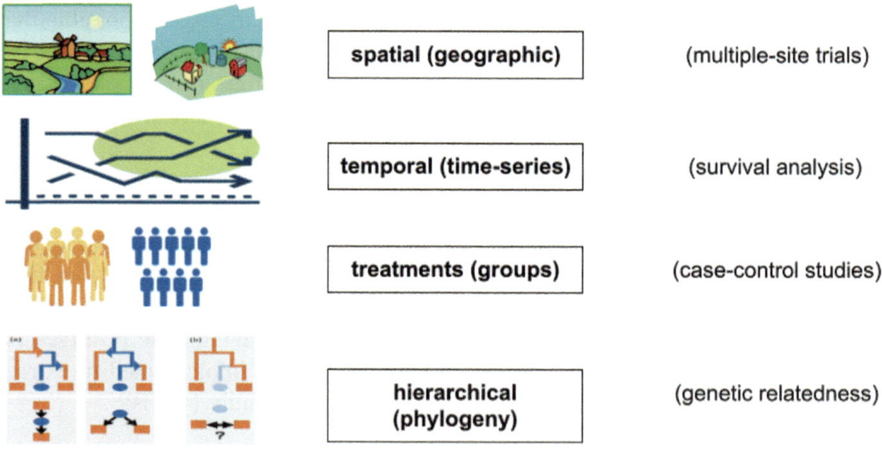

Fig. 3.4 Types of structural dependencies in real-world biological datasets. (Examples are reported within brackets) [cliparts from www.clipartmax.com]

structures are often ignored, resulting in serious underestimation of the predictive error [44]. In addition, most semi-parametric machine-learning methods do not allow for data stratification to be explicitly accounted for in the statistical model of analysis. Hence, validation designs more robust than standard random (uncorrected) cross-validation are needed. This is because when there are ties between the training and validation datasets (i.e. the validation data are close in space, time, treatment, ancestry etc., to the training data), models tend to appear more accurate than they really are (models that overfit the training data are privileged).

Block cross-validation, in which data are split strategically rather than randomly, can address these issues. Building on the work by Roberts et al. [45], and keeping in mind the illustration in Fig. 3.4, we can make the following examples:

(1) **Spatial structure**. We take the example of multi-site agronomic trials: e.g. peach (*Prunus persica*) varieties cultivated in different countries and in different regions within countries, so that a gradient can be imagined (the environmental conditions in geographically close regions will tend to be similar). If, for instance, we want to predict peach yield from genomic and environmental data, random cross-validation will produce an inflated estimate of the predictive ability: a proportion of the training and validation datapoints from nearby locations will be spatially related, and this will violate the requirement for validation data to be independent. A solution is to apply block cross-validation, where blocks are designed across the spatial structure (i.e. in contiguous geographic locations). This forces validation to be performed on data that, on average, are more spatially distant, thereby reducing the spatial dependence and, consequently, also the bias in predictive ability.

(2) **Temporal structure**. We can think of the daily milk yield of cows over successive lactations. We can model milk yield as a function of the cow's breed, age, body weight, the herd to which she belongs and her average feed consumption. However milk yield may also be affected by other covariables that we haven't measured, like the climate, which is also structured in time. Similarly to spatial dependence, here we have data that are correlated in one dimension, i.e. time, and random cross-validation would incur the same problem: training and validation data measured at more or less the same time will be temporally related, thereby "cheating" the validation and yielding inflated estimates of the accuracy. Again, blocks of contiguous time would help ensure some degree of independence between cross-validation folds, reducing the prediction bias (e.g. not using "future" examples at training to then predict the future).

(3) **Experimental/group structure**. In this case the data stratification is caused by some kind of grouping, which can be experimental groups (e.g. treated and control subjects, affected/unaffected individuals) or other grouping variables inherent to the data at hand (e.g. sex, dietary habits). An illustration is provided by racehorses from different teams: we may want to predict the expected racing results of individual horses based on results during training and morphometric measurements. Random cross-validation would ignore other covariables potentially confounded with the racing teams (e.g. trainer, diet, facilities) that could

make observations within the same team more similar and inflate the estimated accuracy of prediction. Blocking by horse team, i.e. using data from other teams to predict horse racing performances in the target team helps circumvent the problem of the underlying stratification and delivers a more realistic estimate of the prediction error.

(4) **Genetic structure**. We provide here a different type of example, where designing cross-validation to explicitly avoid blocking for the data structure helps in correctly addressing the prediction problem. We take the case of whole-genome predictions in animal breeding: here we want to predict the future performance of new generations of animals (offspring, siblings) based on the genome of their ancestors (parents, grandparents). Stronger ties between the training and validation datasets (e.g. parent-offspring relationships, half/full sibs relationships) actually increase the predictive ability of whole-genome predictions [46]. In this example we don't want to use blocking, e.g. by genetic lines, as this would mean using data from specific lines to make predictions on other lines not included in the training data. The objective, however, is usually to use genomic data from older animals to predict the future performance of younger animals (the offspring), or to make predictions on horizontal relatives, i.e. siblings. Therefore, random cross-validation will give an estimate of predictive ability that is closer to the target than block cross-validation by genetic line or family. An even better choice would be to specifically design cross-validation to mimick the target objective, e.g. include siblings in both the training and testing sets.

The blocking strategy must be carefully considered as it holds out entire portions of the predictor space (e.g. geographical locations, time intervals etc.) and this will force the model to predict outside of the ranges of the training data. Block cross-validation (if applicable) is very often more appropriate in principle than random cross-validation for predictions with new data or predictor space. Sometimes, however, dependencies in the data can actually be used to estimate the quantities of interest, as we saw in the case of whole-genome predictions that by design exploit links (relatedness) between the training and validation sets. Predictions from block cross-validation have larger error than those from random splits and an open question is whether this is closer to the "true" error rate. This is probably dependent on the specific problem, and remains to be determined on a case-by-case basis.

3.5 Can We Measure the Performance of Unsupervised Learning?

In Sect. 2.4 (Supervised vs unsupervised learning problems), we saw that unsupervised learning problems in general tend to be more difficult than supervised problems. You have no answers available to learn the rules, so you are left with only one of the ingredients shown in Fig. 2.2: the data, nothing more! Additionally, with no answers there is no obvious way to measure the performance of our models. In this section, we are initially presenting some ad hoc methods to empirically evaluating models for unsupervised learning; then we'll touch upon some more specific metrics and even possible (cross)validation approaches applicable when answers are not available (or are only partially available): for instance, a very simple idea is to split the data and check how reproducible or variable the results are (more specific unsupervised cross-validatory techniques are described in the chapters on **clustering** and **anomaly detection**). We start by taking the example of the popular Principal Component Analysis.

Principal Component Analysis

Principal Component Analysis (PCA) is a statistical method used for unsupervised learning problems. PCA works by summarising a high-dimensional dataset (many numeric features) with a smaller number of representative variables (the "**principal components**"), which together capture most of the variability present in the original data. (Principal components (PC) j are linear combinations of the original p features, weighted by specific coefficients, the *loadings* (ϕ); the *scores* (new feature values) for example i and PC j are: $z_{ij} = \phi_{1j}x_{i1} + \phi_{2j}x_{i2} + \phi_{pj}x_{ip}$. Principal components are centered (mean = 0) and orthogonal one to the other (uncorrelated), and their variance over n individuals (i) is calculated as $\frac{1}{n}\sum_{i=1}^{n} z_{ij}^2$.) The extracted principal components can be used for many purposes, e.g. clustering, feature engineering/extraction and visualizing patterns in the data.

From a (n x p) data matrix **X** (n records, p features; remember, this is unsupervised learning: features only!), we can obtain a maximum of $min(n - 1, p)$ principal components (1 of the n rows—any—is mathematically obtainable from the remaining $n - 1$ rows, since in PCA data are centered, i.e. mean is zero). We then want to use the smallest number of principal components needed to get a good description of the data. The key question is therefore: how many principal components shall we use?

There is no easy answer to this question. One possibility is to consider the proportion of the total variance in the data explained by each principal component: one can either (i) inspect the **scree plot**[9] of the ordered (decreasing) principal components visually, then select them up to where the bend in the curve occurs (**elbow method**); or (ii) select the smallest number of principal components that are required to explain the desired amount of the cumulative variation in the data.

Figure 3.5 shows the results of PCA applied to data from quantitative nuclear magnetic resonance (NMR) spectroscopy-based metabolomics on dog serum and plasma samples [48]: the dataset consisted of 119 metabolomics variables (p) measured on 2069 dogs (n). Hence, the maximum number of principal components

[9] A "scree plot" (from the English word scree = sloping side of a mountain) is the diagram of the eigenvalues from the eigendecomposition of a matrix, in decreasing order [47].

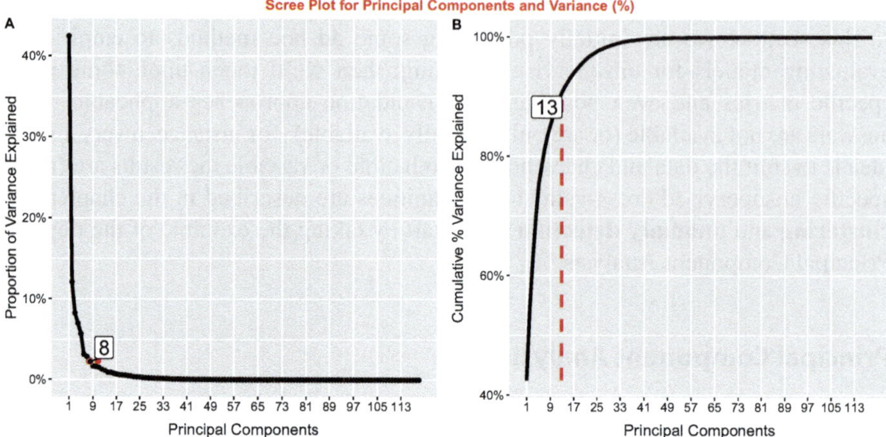

Fig. 3.5 (**a**) Percentage of variance explained by individual principal components, in decreasing order; (**b**) cumulative proportion of the total variance explained by successive components. Data from https://onlinelibrary.wiley.com/doi/10.1111/vcp.12954

is $min(n - 1, p) = 119$. In panel (A) of the figure we see that after 8 principal components the scree plot curve clearly bends, and remains rather flattish afterwards; in panel (B), we highlight the smallest number of principal components (13) that cumulatively account for 90% of the total variance (this threshold is obviously arbitrary, and one can choose whichever number makes sense depending on the specific nature of the problem at hand).

Clustering

A typical unsupervised problem is clustering: given a set of unlabeled data, we want to find homogeneous subgroups of the records, i.e. we want to divide the data into subgroups so that samples within subgroups are similar to each other, while samples across subgroups are dissimilar from each other. The key question is obviously determining what is the best number K of clusters (subgroups).

An interesting approach is that employed by the ADMIXTURE algorithm for the identification of subpopulations from genomic data [49]. From sound "statistics of yore", the elements x_{ij} of a data matrix $\mathbf{X}$ with i rows (samples) and j columns (variables) can be predicted by model-based combinations of parameters of the remaining data [50]. In ADMIXTURE this result is used to predict g_{ij} elements of the matrix $\mathbf{G}$ of i individual genotypes at j genomic loci as:

$$g_{ij} = 2 \sum_{k=1}^{K} q_{ik} P_{kj} \tag{3.3}$$

where P_{kj} is the allele frequency at locus j in subpopulation k and q_{ik} is the proportion of subpopulation k assigned to individual i and computed via a clustering algorithm. Note that **this prediction is a function of** K, the number of clusters that we want to determine!

With this tool in our hands, we can pretend that we don't know some of the genotypes g_{ij} ("*masking*") and predict them with Eq. 3.3: we can then measure the error of prediction and use this to guide us in the process of choosing K. How do we proceed? We: (i) mask a portion of the data; (ii) choose a value of K and cluster the data on the training set (i.e. obtain the q_{ik} values); (iii) predict the masked data with Eq. 3.3 (a function of K!); (iv) measure the prediction error on the set of masked data. We then repeat this process for all the values of K that we want to test and choose the K that gives the **lowest prediction error** (ADMIXTURE actually implements cross-validation to do this: we already provided a detailed description of cross-validation in Sect. 3.2).

Anomaly Detection

A crude fact of life is that there is no dataset without errors, and since we know that **"garbage in" means "garbage out"** (GIGO [51]), we need to get rid of the errors. Easy to say, but we first need to **identify the errors**, and this is precisely what anomaly detection does: finding out which data-points are so **"anomalous"** that we can assume they don't belong to the dataset. Besides data cleaning, anomaly detection has many other important applications: in many real-life situations it is relevant to identify anomalies to trigger actions: e.g. detect credit card frauds so as to then block any transactions made with that card; identify defective components of a machine before it is assembled. As an illustration, let's think of a dairy company that processes, packages and distributes fresh milk: they can use milk lactose, acidity, non-fat solid content, fat, protein and aflatoxin M1 (AM1) as detection indicators of fresh milk samples (example from [52]).

Anomaly detection is typically an unsupervised problem; however, sometimes for a few records we may know which data-points are the errors and which are not (semi-supervised problem). When a sufficient amount of labeled data is available we can use the standard error metrics for classification (e.g. ROC/AUC, precision, recall etc.: see Sect. 3.1 (Error metrics for classification problems)) to measure the performance of our anomaly detection algorithm. Otherwise, alternative criteria must be used that can be computed on unlabeled data: we will mention three such approaches.

The first is the **Mass-Volume (M-V) curve** [53]: we assume that the data come only from the normal class with a probability density function $f(x)$, where x is the input data (e.g. lactose, acidity, fat etc. from our fresh milk example). The lower $f(x)$, the more anomalous the record x: in Fig. 3.6 the red segments show these "anomalous" values of x that have a probability $< q$ of belonging to the distribution of normal samples (unlikely samples). The goal is to find the right value of the threshold q to detect the anomalies correctly.

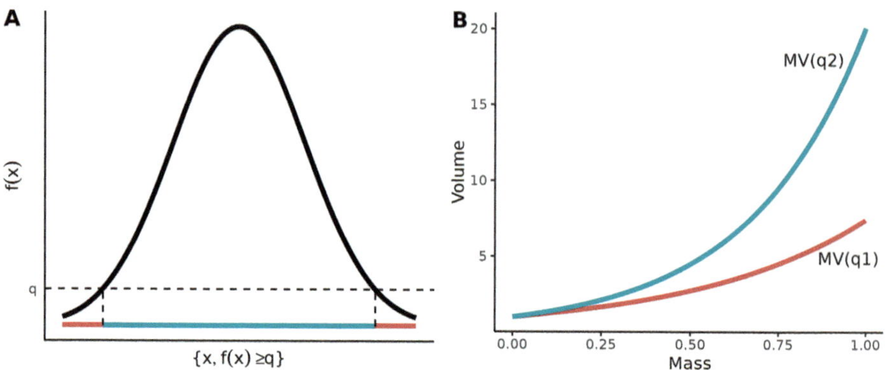

Fig. 3.6 Panel (**a**): probability density function of data x; q is the probability threshold for anomalies (rare events). The red segments indicate anomalous values of the data, while the blue segment indicates "normal" data. Panel (**b**): example for M-V curves for two different values of q. In this case $q1$ results in a lower area under the curve and is to be preferred

From $f(x)$ we can calculate the "mass" (M) and the "volume" (V, defined as the "Lebesgue measure" λ [54]) of the non-anomalous part of the density distribution: $M = P(f(x) \geq q)$ and $V = \lambda\{x, f(x) \geq q\}$. These two quantities can be plotted one in relation to the other, thus defining the M-V curve: both M and V are functions of q, and the value for q that minimises the area under the M-V curve will give us our threshold for anomaly detection![10]

The second approach is based on creating **synthetic records**: from the entire feature space of our data (e.g. every possible combination of lactose, acidity, non-fat solid content, fat etc. from our fresh milk example) random samples are taken and, based on their probability (likely or unlikely samples), they can be labelled as "normal" or "anomalous". Once we generate enough such synthetic records, our anomaly detection algorithm can be evaluated on how accurately it will identify the anomalies from the entire dataset (real + synthetic records together). The tricky thing here is to obtain accurate estimates for the probability of belonging to the feature space of the data, which is a rather difficult exercise [55].

The third approach is known as **cross scoring** and it's a sort of cross-validation based on the idea that similar data from different sources will be somewhat different, and could be considered "anomalous" with respect to each other. Using our dairy company example, we can think of fresh milk produced in different plants (e.g. different geographical regions that use milk from different farms): each plant will have different baseline values for the parameters used to check the quality of fresh milk (lactose, acidity, fat etc.). Let's assume that we have two production plants, "A" and "B". For each plant, we will take a subset of the milk samples to define the baseline values for the quality parameters and leave aside a smaller fraction of

[10] A *Python* implementation of the M-V curve (and the closely related E-M curve) can be found in https://github.com/ngoix/EMMV_benchmarks.

the samples to test the anomaly detection algorithm. Importantly, these test samples will be compared against both baselines: their own, and that from the other dairy plant, each time producing probabilities of belonging or not (being similar or not) to the samples of that specific plant. Some of these test samples will be labelled as belonging to plant "A": these will be considered "normal" samples when applying the anomaly detector to milk from plant "A", and anomalies when analysing milk from plant "B". Now we have our labels (somewhat artificial) and we can measure the performance of anomaly detection!

3.6 Practical Tips for Validation

We discussed the how-to, pros and cons of the most common approaches to the validation of results from predictive models. When moving from theory to practical applications however, we face several operational questions, and decisions need to be made: for instance, how much data is enough data? How should we split the data? What if there are errors in the data? In this section, we want to give you some guidelines to help you put these concepts into practice with your data.

Effect of Dataset Size

The number of available samples influences the performance of any statistical model. In general terms, having a larger dataset is always better (if you want to know more on this, refer to data-driven science and the 4th paradigm [56]). However, getting more and more data sometimes becomes a problem: this can go from being just a minor inconvenience, to being very expensive (storage, runtime, services), to becoming straight-up impossible. It is unfortunately difficult to precisely define what amount of data is "enough", but as a rule of thumb we can say the following:

- **you have too much data** : in this (very lucky) case don't even bother with cross-validation. If you have a good sense that a reasonable fraction (e.g. 20%) of your data contains enough variability to represent the problem at hand, you can just use it for validation
- **you have just enough data** : this is the standard case when a single validation set containing between 10% and 20% of your data is not representative of the complete variability to be expected in the problem. But rotating them through cross-validation (and maybe repeating the whole process several times) gives you a solid estimation of the actual predictive ability of the model
- **you don't have enough data** : it happens, sometimes. Just don't panic. You can push cross-validation to its limit and implement a strategy called *leave-one-out* where the validation set is reduced to a single sample. However, be aware that not all metrics can be computed in this case, i.e. error rates are still well defined while Pearson's correlation is not, since it requires at least two samples. Another option is to use *bootstrapping* [57], where data are resampled with

replacement to produce many slightly different copies of the original data. As an alternative deep learning -but also machine learning in general- can benefit from data augmentation techniques, as described in Sect. 6.2

Test Set vs. Validation Set, and the Need for Nested Cross-validation

So far we have consistently talked about splitting the data in training and validation sets. But sometimes even this approach is not cautious enough, and you may need to split the data into three parts: training set, validation set and test set. In this case we talk about *nested cross-validation*, which goes as follows:

1. hold out a part of your data (e.g. 10%) as the *test set*, which will not be used until the very end
2. with the remaining data do a regular cross-validation, which will have its own training and validation sets internally. If you are evaluating several models (e.g. different deep learning architectures, or different values for a hyperparameter) collect the metric of choice for each candidate and use it to decide which model is the best one for the studied application
3. apply the selected model on the test set you are holding out and obtain an evaluation on completely new data, which will estimate how the model will perform in the real world

Figure 3.7 illustrates this process of splitting the data to train different models (fine-tuning of the hyperparameters, model building), select the best model through (internal) cross-validation, apply this final model on the whole training set, and finally evaluate the model on the hold-out (independent) test set.

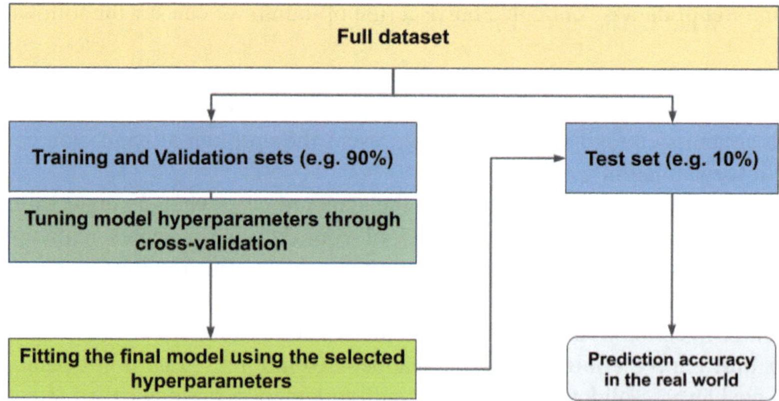

Fig. 3.7 The entire dataset is first split into two partitions. The test set is put aside and will be used at the end to evaluate the performances of the model in the real world. The remaining data is usable for model selection and tuning and is further subdivided by cross-validation or repeated sampling in training and validation sets. The final model, selected based on the metric of choice, is then trained once more on the full tuning data and evaluated on the (hold-out) test set

Regular cross-validation is enough to establish relative performances when comparing different models. However the data used for training cannot be used to extrapolate the model behaviour in absolute terms. A nested cross-validation thus gives the best possible unbiased estimation of the model's performance, since it is evaluated on completely new data never seen during the training phase.

What If There Are Errors in the Data?

Data—especially real-world field data—are not perfect, but are *always* burdened with a *usually small* proportion of errors; these are known as **"noisy" data**, and examples include genotyping errors [58] and swapped labels in case-control studies [59]. Errors in the data arise for a variety of reasons, like instrument or human errors, environmental effects, inherent randomness in the physical process, corruption of the stored data etc.

Errors in the data are known to affect the performance of predictive models negatively [60]. Our research, however, has shown that predictive models are not all equally affected by noise in the data: all methods are robust to small quantities of errors in the data, but local predictive methods, which place higher weight on data in the neighbourhood, seem to be better suited to cope with larger amounts of noisy data, compared to methods that leverage global properties of the dataset [61].

Besides relying on the tolerance of predictive methods, one can take an active role and try to identify and remove, or correct, the errors in the data. This approach is known as **data cleaning**, and it can take many different forms.

The first step in data cleaning is to **identify the (probable) errors**: this is similar to **anomaly detection**, and the techniques discussed in the Section "Anomaly detection" can be used for the purpose. In addition, errors in the data can be identified by using methods based on a model distribution (considering either the fitted values or the residuals) [62], or by looking at scores obtained from a variety of statistical tests for outlying data points like the Chauvenet's and Peirce's criteria or the Grubbs', Dixon's Q, χ^2 and z-score tests [63].

Once probable errors in the data have been identified (with good confidence), one can either go back and **remeasure** (resample) the erroneous data points (if the research budget allows for this), or simply **remove** them (which is a viable option unless there are too many errors and the dataset is "butchered").

Aren't Test and Validation Synonyms?

In plain English the words *test* and *validation* have a somewhat similar meaning, and even in data science different sources mix the notations, so that you may find articles or software that call "test set" what we have defined as "validation set" in the previous sections (and vice versa).

We do not claim to follow the RIGHT WAY on the topic, we simply aim to maintain consistency (sometimes with difficulty). In any case, be advised that different conventions do exist.

Part II

Neural Networks and Deep Learning Decoded

Neural Networks and Deep Learning Decoded 4

We now introduce the protagonist of this book: *mesdames et messieurs, le* ***deep learning***!

We begin with a quick historical excursus of the main milestones that characterised the emergence and development of neural networks in data science. We will then have a peek into the anatomy of a neural network (what it is, what it looks like, how it is made), and see in detail its main components.

We will use a binary classification example (and a logistic regression benchmark) to illustrate the ins and outs of neural network models, starting from a very simple network that will be the elementary constituent for successive, more sophisticated models.

Finally, this chapter will introduce more advanced components of neural network models and will examine methods for solving such models once they have been assembled.

4.1 A Brief History of Neural Networks (and a Bit of Artificial Intelligence)

Humanity has always been fascinated by the very concept of intelligence. We feel that intelligence—leaving aside the elusiveness of its definition—is the main, distinctive trait that separates us as a species from everything else on the planet, and through the centuries a great number of thinkers tried to concisely understand and explain the very nature of the core of our supposed superiority.

In ancient times, the connection between the brain as an organ and intelligence was still to be discovered, with Aristotle famously theorizing, in the fourth century BC, that the brain's function was that of a radiator used to disperse the excess heat generated by the heart [64]. Briefly put, Aristotle believed in the importance of keeping a cool head.

F. Biscarini, N. Nazzicari, *Deep Learning for Life Sciences*, Decoding Evolution 1,
https://doi.org/10.1007/978-3-031-96852-5_4

The connection was eventually discovered, but we had to wait until the twentieth century to obtain a deep enough understanding of the brain's physiology to let us separate the cellular infrastructure from its functionality as a machine able to manipulate information. In engineering terms, it was no earlier than the last century (twentieth century) that we were finally able to separate the hardware from the software.

The first, historical breakthrough in the field of artificial intelligence came in 1943 when Warren McCulloch and Walter Pitts published their paper [65] describing the capability of "all-or-none" networks to implement logical functions, as well as the fact that, given a target logical function, it was possible to find many different networks that implemented it and were thus logically equivalent. This pioneering work allowed for the separation of two fields: on one side the biological research on brain functionality, on the other side the study of neural networks as mathematical abstractions. This second branch eventually generated the deep learning subfield.

But between McCulloch and Pitts's paper and modern neural networks, there were a lot of gaps that needed to be filled. A further breakthrough came from Donald Hebb's work on synaptic plasticity or, in layman terms, the way in which neurons can actually learn things [66]. Neurons are cells connected by elongated structures called synapses. Not all synapses are the same and, in fact, it has been observed that synapses are in constant evolution: they can become thicker, thinner, altogether disappear or be created *ex novo*. The biological details are complicated but the mathematical abstraction is clear: some connections "weigh" more than others. This insight was soon formalized when in 1957 Frank Rosenblatt theorized what became known as the *perceptron* [67] (Fig. 4.1).

But it was not an easy path, nor a straightforward one. During the Sixties, neural networks, together with other technologies all part of the very wide and ill-defined family of *artificial intelligence*, garnered a lot of hype, too much of it, as it turns

> it should be feasible to construct an electronic or electromechanical system which will learn to recognize similarities or identities between patterns of optical, electrical, or tonal information, in a manner which may be closely analogous to the perceptual processes of a biological brain. The proposed system depends on probabilistic rather than deterministic principles for its operation, and gains its reliability from the properties of statistical measurements obtained from large populations of elements. A system which operates according to these principles will be called a perceptron. A model which is

Fig. 4.1 A scan from Rosenblatt's original 1957 report in which the perceptron was theorized for the first time

out. It all came to a halt in the Seventies when the so-called *first AI winter* began,[1] a period of general mistrust towards the real capabilities of predictive machines that led to little-to-no money being put into the field and to a general slowing of progress.

In the Eighties, *expert systems* (computer programmes with hard-coded rules that were used to perform specific tasks with expert-level knowledge) were all the rage, gathering a lot of attention and hope [70]. Neural networks, too, saw big improvements with the invention, in 1986, of the *backpropagation algorithm* [71]. This technique is at the core of the model optimizers still used today[2] and allows the network to adjust the weights of its internal connections so as to minimize the difference between the output of the network and the desired output. As a result the neural network learns.[3]

Just as had already happened in the Seventies, these new technologies and improvements garnered too much attention and hype, to the point where it was believed that human-level general artificial intelligence was very close to being reached. When this failed to live up to expectations a *second AI winter* arrived, dashing all hopes: research programs, investments and careers were upturned, and many scientists had the painful experience of having to change their field of investigation or even their job.

As often happens with technology, the history of neural networks and artificial intelligence is cyclic. Some new model or theoretical breakthrough allows us to solve previously unsolvable problems, a lot of hype arises, and article headlines become more and more detached from the actual capabilities of the tools. The general public really loves to dream, and the idea of building "thinking machines" is always popular.

We are currently living in a third big wave of enthusiasm for neural networks. Some limitations of the backpropagation algorithm (namely the "vanishing gradient problem") have been solved and that, together with additional advancements in computer hardware and data availability, has led to the current era of **deep learning**. As of today, it is not clear where the tide will stop or what will be the technological limit—if any—that will curb the general enthusiasm. Given the rough past of neural networks, the wisest course of action at this point seems to be to keep a long-term optimistic vision without falling into the trap of short-term hype.

[1] When discussing the AI winter, a lot of emphasis is usually bestowed on the 1969 book by Minsky and Papert [68] that stated that perceptrons could not solve the XOR problem, thus contributing significantly to the disenchantment towards neural networks, which were clearly subject of too much hype at the time. The book was wrong, since eight years earlier Frank Rosenblatt had already published a solution to the XOR and similar problems showing that three-layer perceptrons are universal logical operators [69]. Unfortunately, Rosenblatt's work was not really well known at the time. Even more unfortunately, he died in 1971 (aged 43) in a boating accident, thus being prevented from giving further contributions to the field.

[2] See Sect. 4.7.

[3] A more precise expression here would be that the network learns to imitate the data, extracting hidden patterns that are maybe hard to formalize. That the network actually *learns* anything in the common, human sense is an open and debatable question.

4.2 Anatomy of a Neural Network

We mentioned in Sect. 4.1, that a number of technological and theoretical innovations made the practical application of **DNNs** possible. These include **innovations in "scale"**: more powerful **processors** (GPUs, TPUs),[4] massive datasets (**"big data"**: e.g. the 100,000 UK genome project [72]), larger computing infrastructures (e.g. cloud and distributed computing), several flexible and comprehensive programming frameworks (e.g. `Keras/TensorFlow`); and theoretical breakthrough innovations (algorithms for efficiently solving very large DNN models, better designed atomic "ingredients" of neural networks).

But what is a neural network? What does it look like? In Fig. 4.2 (left, top) we saw a simplified sketch of a simple network: basically, we have "neurons" (colored circles) with multiple connections between them (solid black lines). These "neurons" (left, bottom from the same figure) are represented by a circle that receives data in input, processes the data inside the "neuron" (*function approximation*) and produces an output. This structure anatomically resembles that of actual neurons in the human nervous tissue (Fig. 4.2, right): neurons are complexly interconnected to each other, with incoming connections (dendrites) and outgoing connections (axons), and this is why these statistical models have been named neural networks. In deep learning, the "neurons" are usually called **units** or **nodes**, and we'll stick to this non-neuron nomenclature in this book.

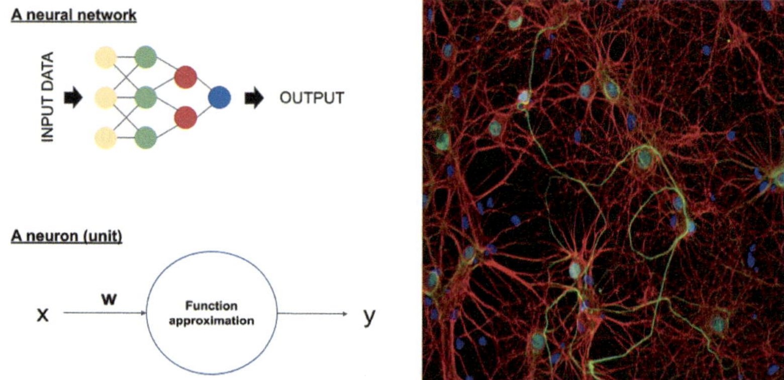

Fig. 4.2 Simplified sketches of: (i) a neural network (left, top); (ii) a neuron (left, bottom); (iii) the human nervous tissue (right). Colored circles represent units (neurons) in the neural network; solid black lines are the connections between units. In the bottom part of the figure, x is the input data, y is the output (result), and w are the weights (coefficients) for the input data (Image of nervous tissue from; https://opentextbc.ca)

[4] GPU: graphics processing unit; TPU: tensor processing unit.

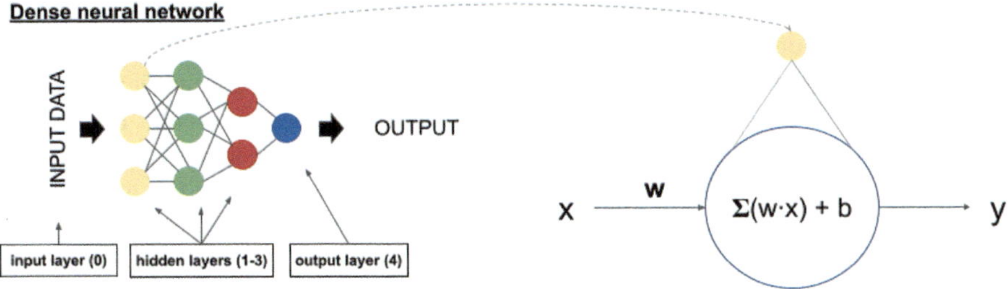

Fig. 4.3 Nomenclature of a dense neural network (left); what happens inside a unit (right)

The nodes (units) lie together in **layers**, and the layers are stacked on top of each other (or side by side) and are interconnected (Fig. 4.3, left). The first layer is usually considered to be the input data (a.k.a. **input layer**) and is indexed at 0 for convenience (layer[0]). The **intermediate** or **hidden** layers are indexed as 1 to $n - 1$: this is where the calculations happen! The last layer is the **output layer**, where the results from the networks' calculations are brought together and the final output is produced (e.g. a number for a regression problems, a label for classification problems). Layers are connected in the direction that goes from the input data to the output results. If there is more than one hidden layer, the network is said to be "deep": these are therefore the much hyped **deep neural networks** (**DNNs**),[5] and the calculations that they do are known as **deep learning**.

When all nodes in a layer are connected to all nodes in the next layer (and so on from one layer to the other in the network), we have a **fully connected** or **dense** neural network (Fig. 4.3, left). This is the most straightforward type of architecture for a DNN.

All right, now we know what a deep neural network is and what it looks like: basically a large number of nodes assembled in layers that are stacked together and interconnected. But what happens inside the nodes?

In Fig. 4.3 (right), we see that the node takes the data **x** as input and calculates a linear combination of the input features (or variables): each feature is multiplied by a **weight** (or coefficient) **w**, then all weighted features are summed together and finally a **bias** term **b** is added.[6] To sum all of this up with the beautifully concise language of mathematics:

$$y = \sum_{i=1}^{m} (w_i \cdot x_i) + b \tag{4.1}$$

where y is the result of the linear combination; x_i is any of the m input features and w_i are the corresponding "weights"; b is the bias term.

[5] The depth of neural networks does not refer to any "profound properties" of these statistical models: it's just the number of layers!

[6] This is the $x \mapsto y$ mapping function approximated by the network in this case.

Now something that is **very important**: if we have only one node (Fig. 4.2, left-bottom), all this turns out to be just a fancy way to perform linear regression, solved through a neural network (actually, a "neural unit") rather than ordinary least squares or maximum likelihood. However, we saw in Fig. 4.3 that we typically have many nodes in a layer and many layers in a network: in a DNN, each node in the first (hidden) layer receives in input all x features and applies Eq. 4.1, i.e. calculates a linear combination of the input features. From each node in layer [1] an output is produced: these are "new variables" which then become the input for each node in the subsequent layer (layer [2]): this is another fully connected layer, where all nodes receive as input all "new variables" and compute another set of linear combinations. And so on for all the layers in the neural network.

Basically, a DNN performs a very **large number of linear regressions!** (Tens, hundreds, thousands, or even hundreds of thousands of linear regression models: Remark 4.2.1). But they won't all do exactly the same thing. Although each node gets the same variables as input, the linear regression models will all be slightly different from each other: in each node different weights will be used for the variables, and some of the variables will not be used at all ($w = 0$). The job of the neural network is exactly to decide which variables to use and how, by **learning the weights**: if it's a DNN, then we call this **deep learning**.

It sounds like the very definition of a complex system: a system composed of a very large number of elementary units (which individually do not necessarily perform a difficult task). And this is just the beginning!

Remark 4.2.1 Equation 4.1 describes the linear regression model that takes place inside one node of a neural network. However in a DNN there is a very large number of nodes, and each node computes a linear combination of the input variables (be these the input data or the output of intermediate layers). A dense DNN therefore amounts to **performing a large number of linear regression models**. For example, a rather smallish DNN with 5 layers of 12 nodes each, plus a final output node has in total $12 \times 5 + 1 = 61$ nodes, and hence will perform **61 different linear regression models**. In each of these 61 linear regression models, the weights w by which the variables are multiplied are learnt independently and are therefore different. Besides, by setting their weights to zero some nodes will not use some variables, which are nonetheless used in other nodes. For real-world applications, much larger DNNs have been developed, with hundreds of thousands or even millions of nodes ([73]): ergo, millions of "linear regressions" (combinations of variables) are calculated!

Function Approximation

In Fig. 2.2, we saw that the objective of deep (statistical) learning is to find the function (the "rules") that map the input data x to the output result y. This function can be (and often is) non-linear and quite complex: the deep neural network will not really find the exact $x \mapsto y$ mapping function, rather

(continued)

it will find an approximation to it. The neural network typically proceeds by successive approximations that are progressively closer to the true (unknown) mapping function, and this is the process known as **function approximation**. Figure 4.4 shows simple examples of function approximation: a true quadratic function and a true cubic function are approximated by a small neural network. As it turns out, neural network models are incredibly good at approximating functions. The **universal approximation theorem** [74] states that neural networks can approximate any function: "no matter what the function, there is guaranteed to be a neural network so that for every possible input, x, the value $f(x)$ (or some close approximation) is output from the network". For practical implementation and examples see the notebook in Sect. A.6

This ability of neural networks to approximate very well the true mapping function for a very wide range of problems, helps explain the roaring success of deep learning: e.g. effectively naming a piece of music based on a short sound sample, identifying cell organelles from microscopy images, predicting a future phenotype (e.g. disease risk) from multi-omics data, translating an Italian text into English (many possible mapping functions, since there are often many acceptable translations of a given text).

This comes with a price, though: being able to learn highly complex and non-linear functions implies that the risk of overfitting the training data (i.e. learning the training data too well) is always around the corner: please refer to Sects. 3.2 and 4.8 for general strategies for minimising the risk of overfitting.

If you wish to learn more on deep neural networks and function approximation, you can check out this excellent online resource: http://neuralnetworksanddeeplearning.com/chap4.html.

4.3 The Building Blocks of Deep Learning

We saw that nodes are the fundamental units of deep neural networks, and that these are first placed side-by-side and then stacked together in multiple layers (the more the layers, the deeper the network). However, we are still missing some essential components: how is the information transmitted through layers? How does the neural network determine the weights (coefficients) for the very large number of parameters typically involved in a DNN model? How does the learning happen?

To answer these questions, we need to examine the basic building blocks of deep learning: (i) **activation functions**; (ii) **cost functions**; (iii) **gradient descent**; (iv) **forward- and back-propagation**.

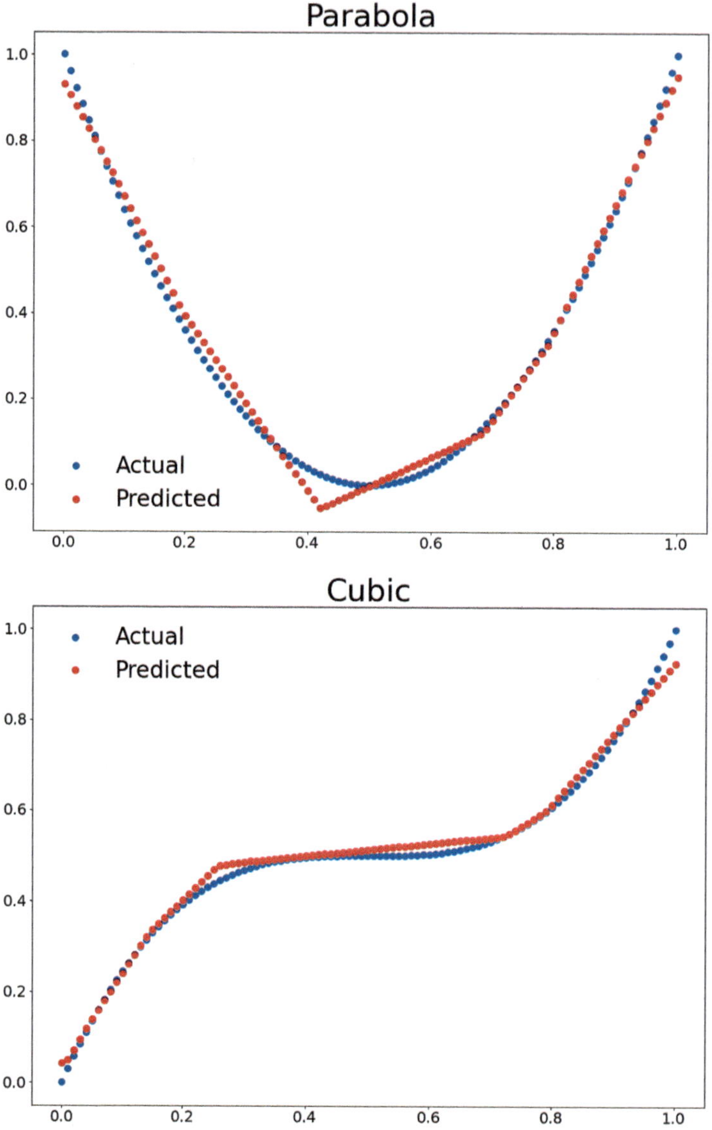

Fig. 4.4 Neural networks can approximate arbitrary functions. In the figure, the same model (2 hidden layers with 10 nodes each) is trained to learn a parabola (top panel) and a cubic function (bottom panel)

Activation Functions

With Fig. 4.3 and Eq. 4.1 we said that the units of a neural network calculate a linear combination of the input variables: but this is not the end of the story. Indeed, the unit (node) receives in input some variables x (the initial input data, or the output of intermediate layers), multiplies these by their respective coefficients w, sums everything up, and finally adds a bias term b. All this produces a continuous numeric value as output, a number in $[-\infty, +\infty]$. However, this number is not passed as it is to the unit(s) in the next layer, but is first "activated" (transformed) by a function; Eq. 4.2 gives the complete picture of the calculations that take place inside the unit: the input variables are combined in the continuous number z which is then transformed by the **activation function** $g(z)$ to produce the output y (Fig. 4.5a):

$$f(x) = \begin{cases} z = \sum wx + b \\ y = g(z) \end{cases} \tag{4.2}$$

All right, we said that the linear combinations of variables are activated by the activation function before they can be further processed: but exactly **when** and **where** do the activation functions come into play? Each time a unit in the network is at work, the input variables are processed and transferred to the next layer through an

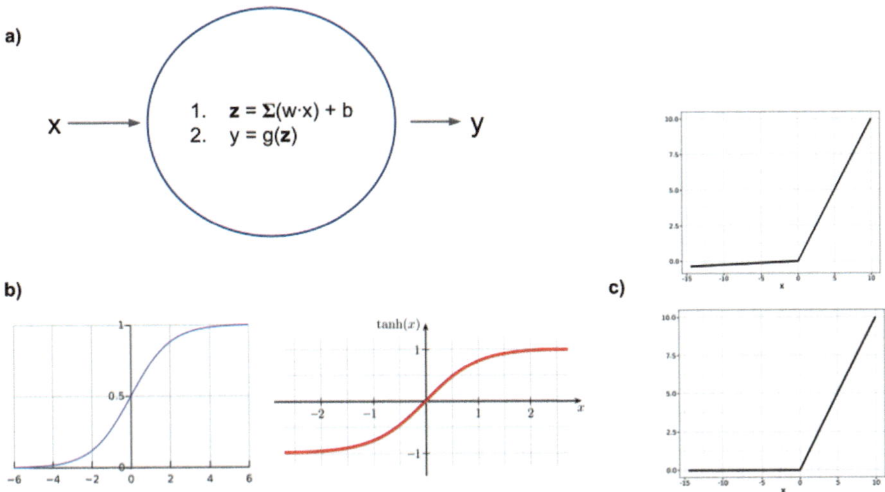

Fig. 4.5 (**a**) Inside the unit, the input variables are first combined together (multiplied by their coefficients, summed up and added to the bias term) and then transformed by the activation function before being outputted and passed on to the subsequent layer in the network; (**b**) sigmoid (left) and tanh (right) activation functions; (**c**) ReLU (bottom) and leaky ReLU (top) activation functions. Sigmoid and Tanh images: Public Domain; tanh image: credit to Geek3, CC BY-SA 3.0 https://creativecommons.org/licenses/by-sa/3.0, via Wikimedia Commons; ReLU and leaky ReLU images from: https://machinelearningmastery.com

activation function, and this happens in all units of all hidden (intermediate) layers and in the output layer. Basically, **activation functions are everywhere** in neural networks!

Now, which are the types of activation function that are used in neural networks? What shape does $g(z)$ have? Here's a list of the most common activations functions:

- Identity function
- Sigmoid (logistic) function
- Hyperbolic tangent function (tanh)
- ReLU (Rectified Linear Unit) function
- Softmax function

Identity Function Let's start with the simplest activation function: the **identity function**, a.k.a. **linear activation function**. In mathematical terms, this is $I(z)$, i.e. it returns the value z that comes from the combination of input features and learned weights (+ the bias term): succinctly, $z = \sum wx + b \rightarrow y = I(z) \rightarrow y = z$ (identity!). In practice, the linear activation function is never used (for good reasons: see the box "Why not linear?" below): the only exception is the output layer in regression problems, where the expected output is exactly the linear combination of input features plus the bias (a continuous value).

Why Not Linear?

We said earlier that the linear (identity) activation function is never used in deep learning (with the exception of the output layer in regression problems): why? This has to do with function approximation. DNNs (deep learning) are excellent at finding complex non-linear relationships in the data (e.g. between features and target variables). However, if we used the identity activation function, the (intermediate) output of each intermediate layer would just be a linear combination of the input data, and so no matter how many intermediate layers we have, the final output $\hat{y}$ would be a linear combination of the initial features $\mathbf{x}$:

$$\begin{cases} \text{layer 1}: h_1 = w_1 \cdot x + b_1 \\ \text{layer 2}: \hat{y} = w_2 \cdot h_1 + b_2 \end{cases}$$

$\rightarrow \hat{y} = w_2(w_1 \cdot x + b_1) + b_2 = (w_2 \cdot w_1)x + (w_2 \cdot b_1 + b_2)$.

Deep learning would just end up being a very expensive way of doing linear regression!

Sigmoid Function The first neural networks waves were dominated by the **logistic activation function**, a.k.a. the **sigmoid function**: this is shown in Fig. 4.5b (left), and converts a continuous input in $[-\infty, +\infty]$ (the by now famous linear combination of features, z) into an output bounded between 0 and 1 (a "probability"). This is done via the following transformation: $\sigma(z_i) = \frac{e^{z_i}}{1+e^{z_i}}$. Despite being historically very popular, the sigmoid activation function has now fallen somewhat out of favor: the reason lies in the "gradients", the slopes of the curve. For extreme (very large or very small) input values, the slopes are virtually flat, which translates to derivatives that are very close to 0. Since the algorithms used to solve the neural network models are based on derivatives, when these are very close to zero the updating of the coefficients (the solving of the model) gets very slow, hence the learning is impractically slow (the derivative of which we speak is actually that of the loss function with respect to the weights, to be exact: through the chain rule, this will inevitably contain the sigmoid function, if this is used as activation function somewhere in the network. Specifically, the derivative of the sigmoid function is: $\sigma'(x) = \sigma(x) \cdot (1 - \sigma(x))$—for details see the accompanying Jupyter notebook with the numerical illustration of backpropagation).

Hyperbolic Tangent Function The rescaling of the sigmoid function leads to the **hyperbolic tangent function** (tanh, Fig. 4.5b (right): $\tanh = 2\sigma(2z) - 1 = \frac{e^{2z}-1}{e^{2z}+1}$. The output of tanh is in $[-1, +1]$, with mean=0, which is actually equivalent to a centering of the data (remember that the output of one layer is the input of the next layer): centered data have desirable numerical properties that bring about a more efficient learning process in the intermediate hidden layers (a similar line of reasoning led to the concept of batch normalization: see the Section "Batch normalization layers"). Still, *tanh* suffers from similar limitations as those of the sigmoid function when z is very large or very small (flat curve, almost zero gradient, slow/virtually impossible learning). Tanh finds application in specialized layers and units, most notably in recurrent neural networks (RNN) which up to now have been at the core of natural processing language (NLP: more on this in Chap. 5).

ReLU Function We now come to one of the major building blocks of deep learning: the **ReLU** activation function. ReLU stands for *rectified linear unit*, and it has been about since the late 1960s in specialized contexts. In 2011 ReLU was found to allow for much faster and more efficient training of neural networks [75], and this greatly contributed to the booming developments of deep learning that have occurred since the second decade of the twenty-first century. ReLU is very simple: $\mathrm{ReLU}(z) = max(0, z)$, which means that the function will return the input itself when this is positive ($z > 0$), and 0 otherwise ($z < 0$) (when z=0, ReLU(z) is 0 in any case). ReLU has a very simple shape (Fig. 4.5c, bottom) and a very simple derivative: $f'(z) = 1$ when the input is positive, 0 otherwise. Table 4.1 summarises the behavior of ReLU and its derivative for different input values. Mathematically, the ReLU function is not differentiable at zero (derivative not defined): however, from an operational standpoint (algorithmically) either 0 or 1 can be chosen as derivative in zero for the practical implementations of this activation function. ReLU displays the following properties: (i) it is easy to compute (both the activation and the derivative); (ii) it gives a sparse representation: many output

Table 4.1 Values of ReLU and its derivative for input (z) larger, smaller or equal to zero

Input	ReLU(z)	f'(z)
z>0	z	1
z<0	0	0
z=0	0	0 or 1[a]

[a]ReLU is not differentiable in 0; operationally, either 0 or 1 can be chosen

values will be exactly 0 (unlike sigmoid and tanh, which tends asymptotically to 0); (iii) it reduces vanishing gradients (there's no gradient slowly approaching zero: it is always exactly 0 or exactly 1). All of this explains why ReLU leads to much faster and more efficient learning, making the training of large multi-layered NNs possible.

Sometimes, ReLU may cause too many units to output zeros: these "flat" zero-slope units may get stuck in the negative side and always output 0 ("dying neurons"). With time, a large part of the neural network may end up doing virtually nothing. The "dying neuron" problem typically occurs when the learning rate is too high or there is a large negative bias term. **Leaky ReLU** is a modified version of the ReLU activation function that, instead of setting negative input values (z) exactly to 0, gives them a very small value, e.g. $0.01z$: hence the activated output value of the unit will be a tiny fraction of the input z, and the derivative will always be the chosen coefficient (0.01 in the example above) [76]. The coefficient for leaky ReLU can either be arbitrarily chosen and set outside of training, or it can be one of the (hyper)parameters chosen by the neural network itself during training.

Softmax Function The sigmoid function converts a continuous input into a single probability value: this is very convenient for binary classification problems, in which once you know the probability of one class, you automatically get also the probability for the other class (as the complement to one). When we have more than two classes, however, this is no longer sufficient and we need to be able to get the probability of each class. This is exactly what the **softmax function** does: it converts input feature values to a distribution of probabilities over the c target classes, where c is the number of classes in the multi-class classification problem. From Eq. 4.3 the similarity between the sigmoid function and the softmax function is apparent; in both functions, negative inputs are converted to non-negative values via exponentiation and probabilities are normalized (sum to 1) by division over the total probability denominator:

$$\text{softmax}(z_i) = \frac{e^{z_i}}{\sum_{i=1}^{c} e^{z_i}} \longleftrightarrow \sigma(z_i) = \frac{e^{z_i}}{1 + e^{z_i}} \tag{4.3}$$

The softmax function is used in the output layer of multinomial (multiclass) classification problems to return the probability of any example to belong to each of the c classes. The softmax function is differentiable, which is convenient for it to be used in backpropagation for the optimization of the weights (the parameters of deep learning models to be "learned").

Cost Functions

We saw in Eq. 4.2 that, inside each unit of the neural network, the input variables are multiplied by their weights (coefficients), they are summed together and then, finally, a bias term is added. The result is then processed by an activation function to produce the output of the unit. We also said that the weights and bias are different for each unit in every layer, which translates into a large number of weights and biases to be learned by the network. The question now is: how are these coefficients estimated (learned)? Remember that this is supervised learning, we have data and answers (Fig. 2.2): at training, we can use the model to make predictions and compare these predictions with the true (available) answers. Initially, these predictions will likely be quite different from the true answers: therefore, we will modify the model (i.e. the coefficients) and get a new set of predictions that will be a bit closer to the true answers. And we will repeat this process as many times as needed until the predictions are satisfactorily close to the true answers. We therefore need a tool to measure how close the predictions and the answers are, i.e. a function that measures the prediction error: this is a **cost function**! Some authors call **loss function** the function that measures the prediction error for a single sample (loss $= f(y_i - \hat{y}_i)$) and **cost function** the sum of the loss over all samples (cost $= \sum_{i=1}^{n} f(y_i - \hat{y}_i)$);[7] in this book, we will use the two terms interchangeably.

Ça va sans dire, there are many different cost functions, depending on the type of statistical problem (e.g. regression vs classification) and the specific implementation. To delve a little more into the details of cost functions, we'll use the example of a **binary classification** problem. Here the answers are '0s' and '1s' (e.g. cases and controls) and predictions are probabilities (in [0, 1]): predictions are obtained, for instance, from a logistic regression model, $p(y = 1|x) = \sigma(\sum \beta x + b)$. Under these premises, we can use the **logistic loss function** (a.k.a. binary cross-entropy) to measure the distance between predictions and true answers. We can distinguish two cases: when the case to predict—y—is '1', and when it is '0':

$$\text{logistic loss}(y, \hat{y}) = \begin{cases} -\log(\hat{y}) & \text{if } y = 1 \\ -\log(1 - \hat{y}) & \text{if } y = 0 \end{cases} \tag{4.4}$$

We see that, when our "answer" is '1' and our prediction (remember: we predict $P(y = 1|x)$, the probability of being '1' given the data) is close to 1, the logarithm of the prediction will be close to 0, and our loss (the prediction error) will be low.

[7] Note that $\hat{y} = g(\sum wx + b)$, and the loss/cost function is therefore a function of the parameters of the model, not of the data.

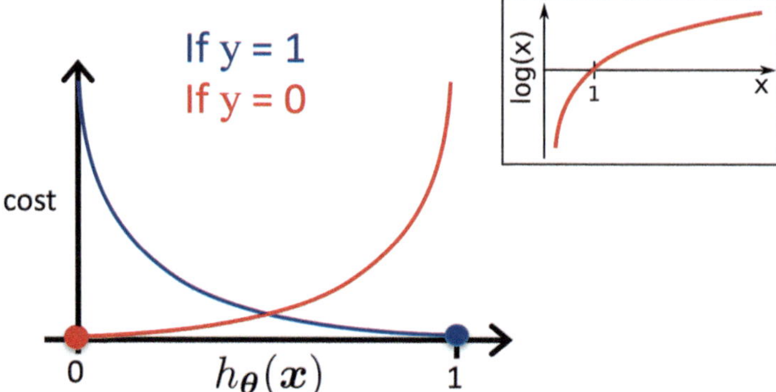

Fig. 4.6 Left: plot of the logistic loss function; on the x-axis, $h_\theta(x)$ is the estimated probability $P(y = 1|x)$ (the activated linear combination of the input features: $\sigma(\sum(x + b)))$; y-axis: cost. Top-right, plot of the logarithmic function: when x is 1, the logarithm is 0; when x tends to zero the logarithm goes exponentially towards $-\infty$ (hence the minus sign before the loss function). Image on the left from: https://medium.com/@jairiidriss/gradient-descent-algorithm-c481c31294a1

Contrariwise, if our predicted probability is close to 0, the logarithm will tend towards $-\infty$ and the loss will be huge. The opposite holds when the answer to be predicted is '0'. Figure 4.6 illustrates this behavior of the logistic loss function.

Summed over all samples, the loss (cost) function can be written in a single expression that combines the two binary cases:

$$L(y, \hat{y}) = -\frac{1}{N}\sum_{i=1}^{N} y_i \cdot log(\hat{y}_i) + (1 - y_i) \cdot log(1 - \hat{y}_i) \tag{4.5}$$

There is also another formulation of Eq. 4.5 that you may encounter in literature (and that is completely equivalent): $L(y, \hat{y}) = -\frac{1}{N}\sum_{i=1}^{N} log(\hat{y}_i)^{y_i} \cdot log(1 - \hat{y}_i)^{(1-\hat{y}_i)}$.

There are many other cost functions available, for both classification and regression problems. Some of these will be introduced later on in more specialized sections. Here we just want to briefly mention **mean squared error** (MSE), which is one of the most common cost functions for regression problems. You are probably already familiar with it: MSE $= \frac{1}{N}\sum_{i=1}^{N}(y_i - \hat{y}_i)^2$.

Remark 4.3.1 In Sect. 3.1 we presented a number of metrics for measuring the performance of predictive models: these are indeed functions, that take as input two arguments (predictions and true answers) and produce as output a quantification of the error (how similar or dissimilar the predictions and the answers are). Why, then, are we now talking of cost functions? They look pretty similar and do more or less the same thing, don't they? Yes, cost functions measure the prediction error, just like the error metrics in Sect. 3.1; however, not all error metrics are suitable as cost functions, as some are discrete or non-differentiable: cost functions must be continuous and **differentiable**! Therefore, what distinguishes cost functions from error metrics are operational aspects: differentiability and efficiency of computation (this is why cost functions are often apparently minor modifications of error metrics, which allow for faster or more efficient calculations).

Gradient Descent

Now we have our tool for measuring how well (or how badly) our deep learning model is doing: the cost function. We want to emphasise that the cost function is a **function of the parameters of the model**: the data are given, and by changing the parameters (the w's and b's in the linear combination z of input variables, see Eq. 4.2) our predictions will get closer to (or farther away from) the available answers (the 0/1 labels, in our logistic regression example). Intuitively, in Fig. 4.6 we see that we want to find the combination of parameters that will give us the lowest possible "cost", i.e. we want to **find the minimum of the cost function**. How can we minimise the cost function? Simply put, just as we minimise any other function: **by taking derivatives**! (Hence, the requirement that the cost function be differentiable)

Minimising the cost function is central to deep learning (and machine learning in general): alternatively, linear or non-linear least squares, or the maximization of a likelihood function can be used to solve statistical models, but these approaches do not find place in deep learning.

Gradient descent is a general algorithm used for the minimisation of the cost function. In its minimal formulation, gradient descent is very simple:

1. (initialisation): start with initial values for the model parameters (the vectors of weights $\mathbf{w}$ and of bias terms $\mathbf{b}$)
2. (descent): change the values of $\mathbf{w}$ in the direction of reducing the cost function ($L(y, \hat{y})$)
3. (minimisation): stop when the minimum cost is reached

This is illustrated in Fig. 4.7b: starting from a point in the space of parameters (the yellow dot in the top corner: initialisation), the algorithm will look around, find the steepest direction and then take a step downward. There are two questions that we need to address: (i) how can we find the steepest direction around a given point in the space of parameters (m-dimensional curve)? (ii) how big is the step that we then take downward? The first question is related to what we already said about the minimisation of the cost function: we take derivatives! The derivative will give us the slope of the function (curve) at the chosen point, thus the largest value is the steepest direction. The size of the step is then given by the learning rate α, a hyperparameter that controls the speed of learning and whose value will be fine-tuned through cross-validation (see previous and later chapters of the book).

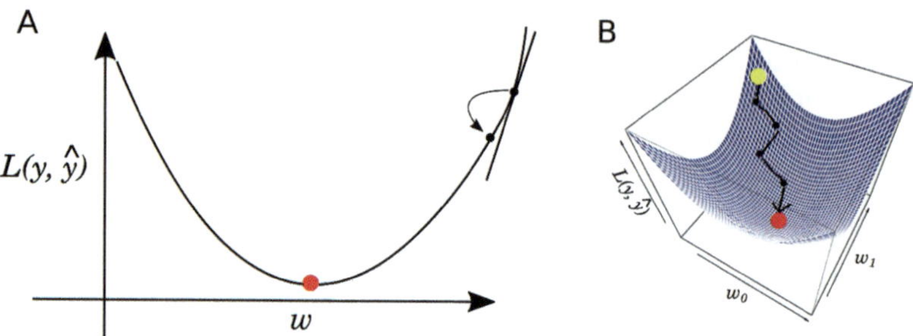

Fig. 4.7 (**a**) 2D-plot of gradient descent, with a single model parameter (w) on the x-axis and the cost function on the y-axis. Remember that $\hat{y}$ depends on w. The tangent line represents the gradient of the cost function calculated for a specific value of w. (**b**) 3D surface of cost as a function of two model parameters (w_1 and w_2). Here we show a learning trajectory, where the yellow dot represents the vector of parameter initializations. Each subsequent black dot forms the path from initialization to the minimum of the cost function (the red dot)

In mathematical notation, considering only one parameter (weight) w_j, we have:

$$w_{j:t} := w_{j:t-1} - \alpha \frac{\partial}{\partial w_j} L(y, \hat{y}) \tag{4.6}$$

where the new (updated) value for the parameter w_j at time t is given by the current value of w_j at time $t-1$ minus the partial derivative of the loss function with respect to w_j times the learning rate α. The reason for the "minus" sign can be seen from Fig. 4.7a: if the slope is positive, moving downward towards the reduction of the loss function means making the value of w_j smaller (subtract the positive slope times the learning rate); if, on the other hand, the slope is negative, reducing the loss function implies adding the slope (times the learning rate) to increase the value of w_j. In both cases, the value of the slope needs to be reversed, hence the negative sign.

A few final considerations on gradient descent. The **learning rate** is an important hyperparameter to fine-tune: if it's too small, the descent will be very slow, with negative impact on the training of the neural network. If the learning rate is too large, the value of the parameters **w** will jump between extreme values (**overshooting**) and the algorithm will fail to reach the minimum of the loss function. Another potential issue with gradient descent is the possibility that the algorithm gets stuck at a **local minimum** of the loss function instead of finding **the global minimum** of the function, which would reflect the best values for the parameters (those that truly minimise the loss). A workaround is offered by the addition of **momentum** to the algorithm (gradient descent with momentum): momentum is an extension of the optimization algorithm that adds a certain amount of "inertia" in a given direction along the space of the parameters, and this allows to go "uphill" on the loss function curve and—if the inertia is sufficient—beyond the local minimum (and towards a global minimum).

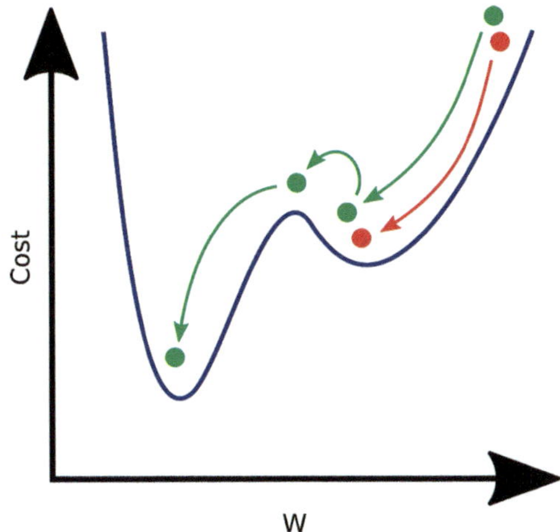

Fig. 4.8 2D profile of the cost function: for different values of the parameter w (weight) the cost changes, with one local minimum and one global minimum. The red dots and arrow show the minimization with standard gradient descent, which gets stuck at the local minimum. The green dots and arrows show gradient descent with momentum: starting from the same initialization, the optimizer is now able to go beyond the local minimum and find the global minimum

Momentum is an additional hyperparameter, which we denote with the letter β, and which is defined in the range $[0, 1]$. It can be understood as adding inertia to the learning trajectory, which allows to jump over a local minimum and progress to a better model optimization (see Fig. 4.8 for a graphical illustration). It is implemented in the calculation of exponentially weighted moving averages (EWMA) of the parameters w to be updated (learned) by the neural network. Intuitively, a moving average takes into account (all) previous values of a sequence of numbers, like the sequence of all previously updated values of the model parameter w. We indicate the moving average by the letter v (for *velocity*, following the metaphor of inertia and kinetic momentum). Being a moving average, its value changes with each updating: the latest value in the sequence (w_{t-1}) and the current moving average (v_{t-1}) are combined to form the new moving average (v_t). The relative weight of these two terms is given by the momentum β. Mathematically, it has the following form:

$$v_t = \beta v_{t-1} + (1 - \beta)\frac{\partial}{\partial w_{t-1}} L(y, \hat{y})$$

which changes the gradient descent updating as follows:

$$w_{j:t} :- w_{j:t-1} - \alpha v_t \tag{4.7}$$

We see immediately that, if $\beta = 0$ (momentum is zero), Eq. 4.7 reverts back to gradient descent without momentum. If $\beta > 0$ (up to the maximum possible value of just below 1), weight is given to the moving average of previous values of the model parameter w. To sum up, gradient descent with momentum is faster and reduces the probability of getting stuck in local minima.

Forward and Back Propagation

Forward propagation and back propagation are the "outer engines" of neural networks (with the units being the "inner engines", in this metaphor): with the given values of the weights vector **w**, forward propagation calculates linear combinations of the input data (activated by the chosen activation functions), sequentially through the layers of the network, until the final output is produced (a continuous value in regression problems, a probability of belonging to a class in classification problems). Back propagation goes the opposite way: once the cost function has been applied to the current network output, the weights will be updated backwards—from cost to output, to activations, to linear combinations of inputs, to weights—using partial derivatives (see Fig. 4.9).

This act of sweeping through the network, once from left to right to produce the output, once from right to left to update the weights, is one **epoch** (a "cycle"—or "iteration"—of network calculations). Typically, a neural network is trained for hundreds of epochs, i.e. cycles of weights updatings through which the network learns to solve the problem (reminder: the weights are the parameters of the neural network model). A detailed description of forward and back propagation, with numerical illustration of the calculations involved, is provided in the accompanying notebook in Sect. A.5

Remark 4.3.2 We want to stress an important aspect here: the difference between **parameters** and **hyperparameters**. This is crucial in any statistical or machine learning model, particularly so in deep learning models. We saw already that in a typical neural network there's a humongous number of "knobs and levers" to tweak: the model coefficients (w, weights), the bias terms, the type of activation function, the number of units, the number of layers ... and many more will be

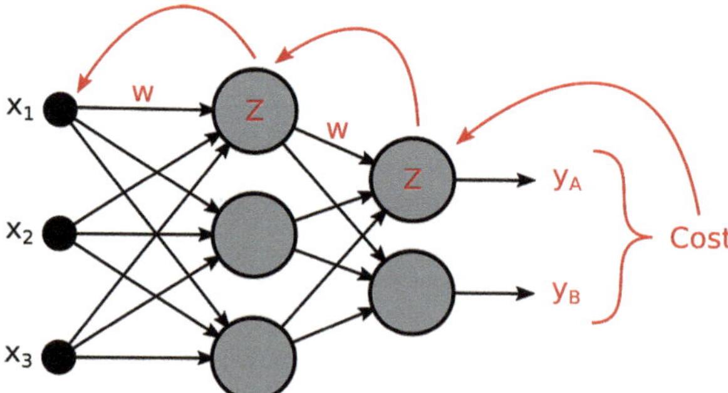

Fig. 4.9 Simple network with forward and backpropagation. Forward propagation is illustrated by the black arrows, from left to right. It computes the network output y based on the input variables x and the current values for the weights w. Backpropagation is illustrated by the red arrows, from right to left (partially: limited to one path over one output, one unit in the output layer, one unit in the hidden (intermediate) layer, one input variable). It updates the current values of the weights w. Forward propagation + backpropagation constitutes an epoch

introduced in the subsequent sections of this book! Some such "knobs and levers" are **model parameters**: these are basically the model coefficients (weights and bias). Others are **model hyperparameters**: the network architecture (units, layers), the activation functions, the cost functions, the optimizer, etc. **Parameters** are **automatically learned** by the neural network during training (cycles of forward and back propagation—the epochs!—on the training data); **hyperparameters** are not learned, but are **fine-tuned** outside of training, typically through cross-validation schemes that may be somehow semi-automatic (more on this in Part 2 "Making it work": Chap. 6).

Parameter Initialisation

Let's take the first forward propagation (first half of the first epoch): we give some values to the model parameters (weights and biases) and then calculate the output of the network. This output is used to calculate the value of the cost function in the current (first) epoch, which is the starting point for back propagation where derivatives are calculated and parameters are updated. The question is: how do we choose these **initial values** for the model parameters?

One option could be to **start with zeros**: we set all weights (and biases) at zero and begin with cycles of forward and back propagation. Would this be a good idea? Unfortunately not. If all weights are initialised at zero, the linear combinations in all neurons will be zero ($z = 0 \cdot x_1 + 0 \cdot x_2 + \ldots + 0 \cdot x_p = 0$). It's easy to see that all ReLU activations will also be zero (ReLU at 0 is not defined, but if we choose either to evaluate it to zero or z, it will still be zero (from above: $z = 0$)). In a regression problem, this means that the final prediction will also be zero, thus nothing will be subtracted from the current cost function value and there will be no learning. In a logistic problem things won't be much different: the sigmoid activation of zero will always produce 0.5, for all examples and from all neurons. This means that all neurons will have the same effect on the total cost of the model (complete symmetry), which will yield exactly identical gradients (derivatives) and exactly identical updates of the weights. At the end of the first epoch we will be "back to square one", with the same weights for all neurons. And so on, indefinitely. Again, no learning. This **rules out zero initialization** as a viable option.

If we think about it, any constant initialization (also different from zero, e.g. one) will lead to the very same problem: for a given constant k, the output of all intermediate neurons will be $ReLU(k \cdot x_1 + k \cdot x_2 + \ldots + k \cdot x_p)$, leading to the exact same influence on the total cost, the exact same gradients, the exact same updates. Therefore, we can also **rule out constant initialization**.

What then? We need to break the symmetry of constant initialization. One obvious option is to use random numeric values as the initial weights of the network. When we do this, however, we need to avoid two potential problems: (1) if the initial values are too large, the weights can start to diverge progressively, getting larger and larger in absolute values (**exploding gradients**); (2) if the initial values are too small, the weights will get smaller and smaller, leading to very slow learning (and approximating the vanishing gradients problem: more on this when we'll discuss RNN).

To illustrate this point, we take the ultra-simplified example of a (2, 2) matrix of weights **W** (2 input variables, 2 units), and we assume that the weights are initialised to different values in one layer, but to the same values over all L intermediate layers (we ignore the output layer in this example). We further assume that the linear activation function is used in all units. Under these conditions, the final matrix of weights reduces to $\mathbf{W}^{[last]} = \mathbf{W}^{[L]} \cdot \mathbf{W}^{[L-1]} \cdot \ldots \cdot \mathbf{W}^{[2]} \cdot \mathbf{W}^{[1]} = \mathbf{W}^{L}$ (i.e. **W** to the L_{th} power), which is still a (2, 2) matrix. We can easily see that, if the initial values are large, they will grow exponentially (exploding gradients and overshooting); if, on the other hand, the initial values are (much) smaller than one (in absolute value), they will decrease exponentially (slow learning, $\sim$ vanishing gradients).

Intuitively, a better approach will be to pick the **initial values to be different in every unit and in every layer**, with some weights initialised to positive values, some to negative values. This can be more formally described by constraining the randomly picked weights to have mean approximately equal to 0 and variance approximately constant in all layers. Mathematics gives us a beautiful solution to this problem, in the form of the *Gaussian distribution*: the initial values for the weights will be randomly sampled from a normal distribution with mean = 0 and variance inversely proportional to the number of units in the layer.

Remark 4.3.3 As you can easily imagine, besides randomly sampling initial values from a Gaussian distribution, several more **advanced techniques for parameter initialization** have been proposed (reviewed by Narkhede et al. [77]). A few are listed below:

- **Xavier-Glorot-initialization** is a technique proposed by *Xavier Glorot* and *Yoshua Bengio* that scales the initial random weight values by the square root of the number of inputs and outputs (connections: n_{in}, n_{out}) of the layer: multiplicative scale $= \sqrt{\frac{2}{n_{in}+n_{out}}}$. This works well for networks that use the sigmoid or tanh activation functions, and helps them avoid too quick saturation (the reaching of asymptotic activation values, which then slow down or even stop gradient descent and learning)
- **He-Kaiming-initialization** (a.k.a. Kaiming) is a modification of the Xavier initialization that scales the initial weights by the square root of two over the number of inputs: multiplicative scale $= \sqrt{\frac{2}{n_{in}}}$. It works well with networks that use the ReLU activation
- **Orthogonal initialization** is a technique that (softly) constrains the matrix of initial weights to be orthogonal (rows and columns are mutually perpendicular: their dot product is zero) and to have unit norm (they sum to one: orthonormality). This prevents activations and gradients from becoming too correlated. Orthogonal initialization works well with recurrent neural networks (RNNs) that have long-term dependencies. Orthogonal initialization requires the weight matrix to be square, and may therefore not be applicable to layers that have different numbers of outputs and inputs

Vanishing Gradients

When describing the activation functions in Sect. 4.3 above, we said that the once-popular sigmoid activation function has now lost favor, because for extreme input values the slopes are virtually flat, and this translates to derivatives that are very close to 0. This can be visually appreciated in Fig. 4.5b (left).

Why is this relevant? We saw that the neural network learns the model parameters by minimising the cost function via an algorithm (optimizer) like **gradient descent**. Gradient descent is applied to the neural network through the **backpropagation algorithm**, which takes into account all connections in the network and propagates derivatives backwards across the layers. You can think of the neural network as a series of nested functions (something like $u(g(\ldots f(x)\ldots))$, and backpropagation is the way to "crack" this complex derivative. More details on backpropagation are in the accompanying notebook in Sect. A.5

If the sigmoid function is used to activate units in the intermediate layers, this means that we need to take its derivative. The derivative of the sigmoid function $\sigma(x)$ is: $\sigma(x)' = \sigma(x) \cdot (1 - \sigma(x))$, which contains the sigmoid function. For very large (or very small) input values this evaluates asymptotically to 0 ($\sim 1 \cdot \sim 0$—or the other way around— $= \sim 0$). Since this is a component of the chain rule for calculating the values of the vector of weights **w**, if it's zero, or close to zero, so will be **w**. If the sigmoid function is used in many units of several intermediate layers, and the calculations are repeated over many epochs (cycles of forward and back propagation), it is likely that many weights will tend to be equal—or close—to zero. This is what we call **vanishing gradients**, and causes the network to learn very slowly or not be able to learn at all.

We introduced vanishing gradients in the case of the sigmoid activation function: in Sect. 5.5 we'll see vanishing gradients in a different setting, that of recurrent neural networks (RNN).

4.4 Logistic Regression from a Neural Network Perspective: Binomial and Multiclass Classification Problems

In the previous sections, we already took—more or less explicitly—the perspective of logistic regression to illustrate the basic functioning of a neural network: e.g. when discussing cost functions and gradient descent in the Section "Cost functions". As a matter of fact, from Fig. 4.5 (part A) we can see that if we replace the generic activation function $g(z)$ with $\sigma(z)$ (the sigmoid activation function) we have a one-unit neural network that is exactly equivalent to a logistic regression model for binary classification. This is sketched in a tad more detail below in Fig. 4.10 (a

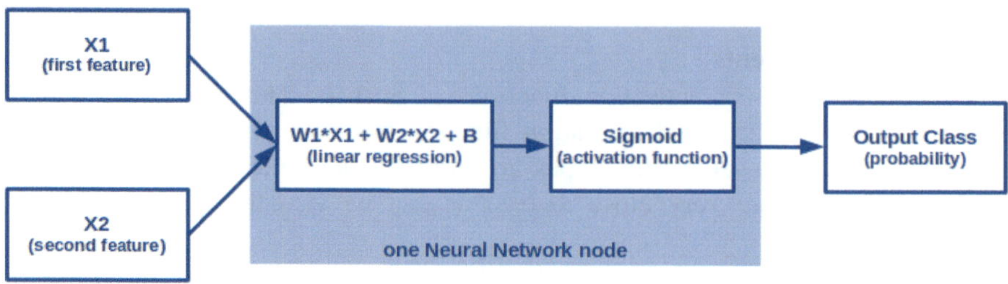

Fig. 4.10 One-unit neural network for binary classification. Two input features, x_1 and x_2, enter the unit and are: (i) linearly combined by multiplication with their coefficients w_1 and w_2 and addition of the bias term b; (ii) "activated" by the sigmoid function into $P(y = 1|x)$ (probability of belonging to class '1'). The output is then class '1' or '0', depending on the chosen probability threshold (usually 0.5)

worked-out example of logistic regression with neural networks can be found in the accompanying notebook in Sect. A.7).

A natural extension of logistic regression for binomial classification is multinomial logistic regression for multi-class classification, which is known as **softmax regression**: in the Section "Activation functions" we introduced the **softmax (activation) function**, which is a generalization of the sigmoid function and, instead of a single probability value, returns a probability distribution over the c target classes (where c is the number of classes in the multi-class classification problem: see Eq. 4.3).

In Fig. 4.11, we sketch the anatomy of a one-layer neural network model for multiclass classification: this single layer (a.k.a. *softmax layer*) has as many units as there are classes (three in this example). It is a fully connected (dense) layer, i.e. all units in the softmax layer receive as input all features in the dataset, and perform independent linear combinations of the data. For illustration, we take the famous *Iris* flower dataset [78], used by Ronald Fisher in his paper on linear discriminant analysis. This dataset contains petal and sepal measurements for three *Iris* species: *Iris setosa*, *Iris versicolor* and *Iris virginica*. It is therefore a three-class classification problem for which we can use softmax regression. (A worked-out example of softmax regression with neural networks on the *Iris* flower dataset can be found in the accompanying notebook in Sect. A.8.)

To solve softmax regression with backpropagation, we need to define the appropriate loss function: this is the **categorical cross-entropy** function, which again is a generalization of the logistic loss function (a.k.a. binary cross-entropy function Eq. 4.4) for binary classification (Section "Cost functions").

The categorical cross-entropy function has the following form:

$$\mathrm{L}(\mathbf{y}, \hat{\mathbf{y}}) = -\sum_{j=1}^{c} y_j \cdot \log(\hat{y}_j) \tag{4.8}$$

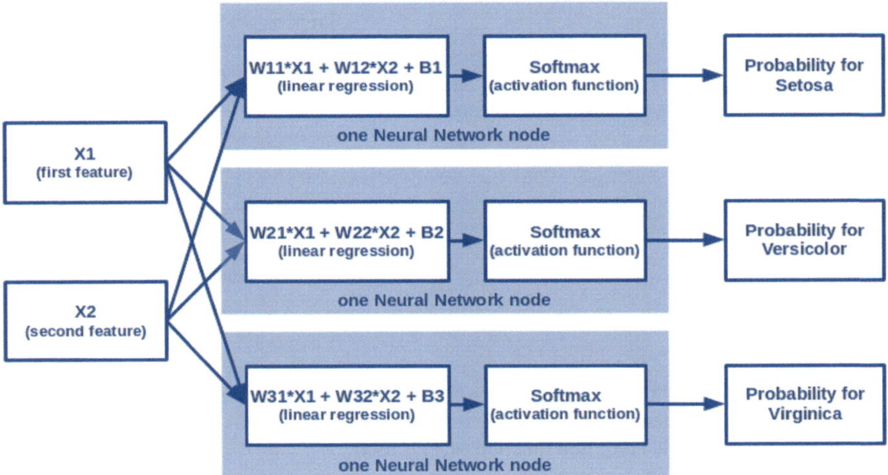

Fig. 4.11 One-layer neural network for multiclass classification (**softmax regression**). This single layer has as many units as there are classes (three in this illustration), each performing a linear combination of the input features and softmax activation. Each unit will use a different set of parameters: w_{lm}, b_l, where m is the feature and l is the layer. For instance, with input features, x_1 and x_2, the parameters will be $w_{11} \cdot x_1 + w_{12}x_2 + b_1$, then $w_{21}x_1 + w_{22}x_2 + b_2$, and so on. For each sample in the data, the output is a vector of probabilities, one for each class, summing up to one

where $\mathrm{L}(\mathbf{y}, \hat{\mathbf{y}})$ is the loss for a single sample, $\mathbf{y}$ and $\hat{\mathbf{y}}$ are the vectors of length c (number of classes) with, respectively, one-hot encoded labels and predicted class probabilities for the sample. When we sum loss values over all samples in the dataset, we get the overall cost of the model in the current epoch.

4.5 From Logistic/Softmax Regression to Neural Networks

In the previous section, we saw how to frame logistic regression, softmax regression (and linear regression) as **neural network models**. These were very simple networks with only the output layer (one single unit overall or per class), and are completely equivalent to the standard logistic/softmax/linear regression, with the only difference being that they are solved through backpropagation instead of maximum likelihood (or least squares etc.).

We now move from these basic implementations to the development of actual **neural networks with multiple layers**, which will be able to express the computational power of deep learning. In Fig. 4.12 we have a two-layer neural network for binary classification: one hidden layer that receives and processes the input data, and one output layer that calculates the probability of belonging to class "1". We introduce a little notation: l indicates the layer ($l = 1 \rightarrow$ first hidden layer), u_u indicates the units in layer l, $\mathbf{A}^{[l]}$ is the matrix of activation values in layer l.

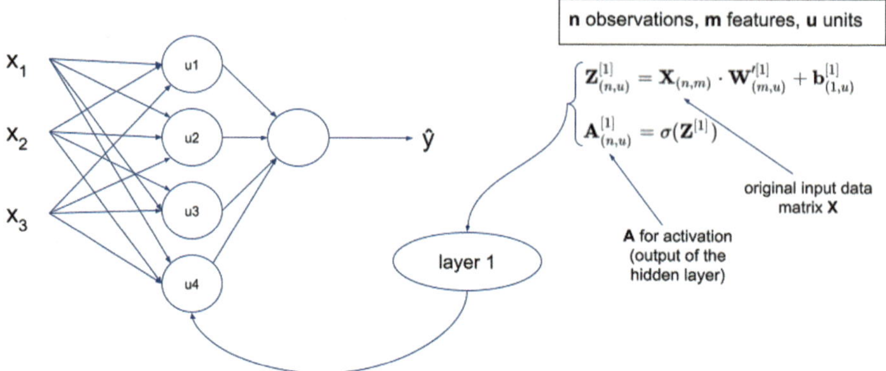

Fig. 4.12 A two-layer neural network: layer 1 (hidden layer) has 4 units u_u and receives in input 3 features x_i. The sigmoid activation function ($\sigma()$) is used

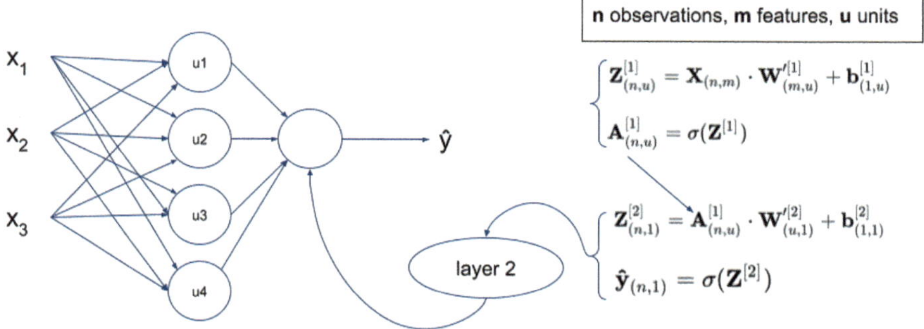

Fig. 4.13 A two-layer neural network: layer 2 receives as input the activation values from layer 1 and calculates the predictions $\hat{y}$

From Fig. 4.12 we see that, in this example, the sigmoid activation function ($\sigma()$) is used also in the hidden layer ($l = 1$):[8] the (n x m) input matrix $\mathbf{X}$ (n examples, m features) is multiplied by the matrix of weights $\mathbf{W}^{[1]}_{\mathbf{m,u}}$ and summed with the vector of bias terms $\mathbf{b}^{[1]}_{\mathbf{1,u}}$ to produce the matrix of linear combinations $\mathbf{Z}^{[1]}_{\mathbf{n,u}}$; this is then activated to give the (n x u) matrix $\mathbf{A}$ of activation values in [0, 1] (the output of the sigmoid function).

The matrix of activations $\mathbf{A}^{[1]}_{(\mathbf{n,u})}$ from layer 1 is the input for layer 2 (Fig. 4.13): in this example, layer 2 is a single (output) unit where the (n x u) input matrix is multiplied by a new set of weights $\mathbf{W}^{[2]}_{\mathbf{u},1}$ and summed with a specific bias term $\mathbf{b}^{[2]}_{\mathbf{1,1}}$

[8] What about vanishing gradients? With only one hidden layer, there's not much risk for the gradients to get too close to zero, and for the learning process to become too slow as a consequence. Still, ReLU would be a valid alternative also in this case.

to produce the matrix of linear combinations $\mathbf{Z}^{[2]}_{\mathbf{n,1}}$, which is then activated to give the (n × 1) vector of predictions $\hat{\mathbf{y}}$ ($P(y = 1|x)$, probability of belonging to class '1', for each sample in the dataset).

This illustration showed a very simple example of a neural network with only one hidden layer and two layers overall (hidden + output): of course, the trick to deep learning is to stack together multiple (many) hidden layers, each with relatively few nodes, to form a **deep neural network** (**DNN**). (Worked-out examples of how to move from logistic/softmax regression to deeper neural network models are provided in the accompanying notebook in Sect. A.10.)

In a DNN, the connection of the activation values in $\mathbf{A}^{[1]}$ with the initial input data is progressively lost, still the network is able to learn the problem at hand very effectively. It is speculated that this learning ability is somehow linked with the compositional representation of the information typical of the human brain [79]: in simple terms, each layer of the DNN focuses on one aspect of the data, which are then combined to get the final result (compositional representation). For example, in a neural network model for face recognition, one layer would learn edges, another layer would learn pieces of faces and yet another layer would learn contours. In a model for speech recognition, different layers would specialize in learning high/low soundwaves, combinations of soundwaves into phonemes, combination of phonemes into words and how to assemble words into sentences.

This is one reason why it is usually better to build a deep neural network rather than a network with one single very wide layer [80,81,82]. Depth is computationally more efficient than width, as shallow NN require exponentially more hidden units (nodes) compared to deep NN: deep NN are $O(log(n))$, while shallow NN are $O(n^2)$. DNN can supposedly learn functions that cannot be learned by shallow networks.

4.6 More Building Blocks of Deep Learning

A Typical (?) CNN (Convolutional Neural Network)

In the following pages we'll examine several more building blocks that are used in common convolutional architectures, namely convolutional, dropout, flatten and batch normalization layers. Many of details can vary and are continuously being updated as deep learning theory is improved, but you can refer to Fig. 4.14 as a general map for the next steps and not get lost in the details.

The type of architecture we are presenting starts with a number of convolutional layers, typically alternated to (max)pooling layers. Then a flatten layer serves as the border between the convolutional part and the dense, "traditional" part of the network. The output of the network is managed exactly as we have already seen when talking about dense networks. In the case of classification problems the very last layer will be a softmax with as many nodes as classes.

Notice that the data shape changes. The face of the data cuboid, which initially corresponds to the height and width of the image, shrinks through the network. On

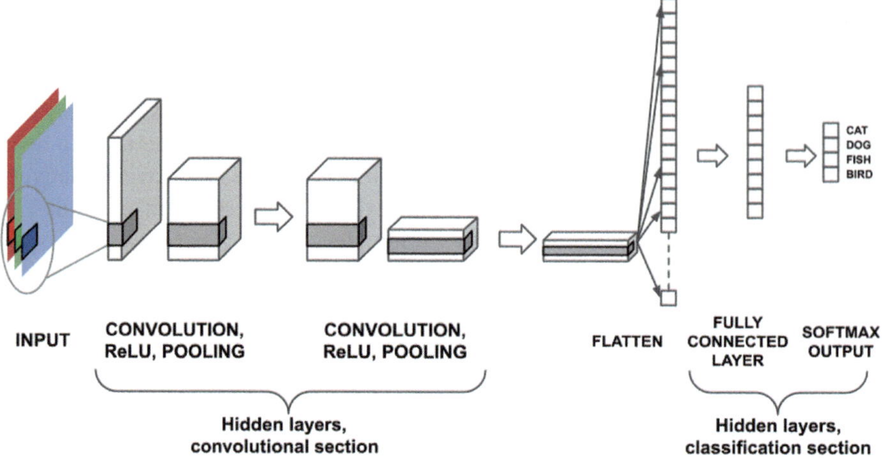

Fig. 4.14 A typical convolutional neural network (CNN) architecture. In real life applications the convolutional section usually has many more layers, while the classification section does not need to be much deeper than what is shown here

the contrary the number of channels starts from three (corresponding to red, green and blue in an RGB image) and grows, creating a very thick data shape just before the flatten layer.

Let's now examine these new building blocks. But first:

What Is a Convolution?

A convolution is a mathematical operation carried out between two matrices. Its symbol is an asterisk: *
Usually, one matrix is bigger (wider, larger) than the other. This bigger matrix represents the pixels of an image, which is being convolved and thus analyzed. The other matrix, called "a filter", is usually very small and represents the type of convolution that is happening. For reference, take a look at Fig. 4.15 which shows the effect of convolving the same image with two different filters.

From the mathematical point of view a convolution is quite straightforward. Given two matrices, the core convolution step is a cell-by-cell multiplication, followed by a big final sum:

$$\begin{bmatrix} A & B \\ C & D \end{bmatrix} * \begin{bmatrix} \alpha & \beta \\ \gamma & \delta \end{bmatrix} = A\alpha + B\beta + C\gamma + D\delta$$

A numeric example (so you can check our arithmetics):

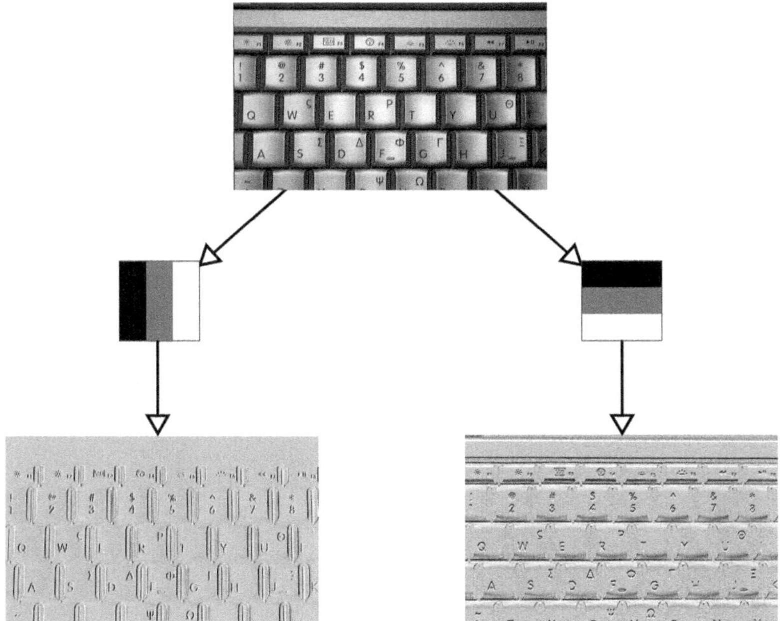

Fig. 4.15 An example of convolution with filters designed to highlight vertical elements (left) or horizontal ones (right). The filters are represented out of scale—their actual size is 3×3 pixels. For reference, the keyboard picture is 1761×1056 pixels. Detail taken from public domain image "Greek Keyboard (Macbook Pro)" (Klilikakis https://commons.wikimedia.org/wiki/File:Greek_Keyboard_(Macbook_Pro).jpg)

$$\begin{bmatrix} 7 & 5 \\ 2 & 9 \end{bmatrix} * \begin{bmatrix} 0.5 & 2 \\ 1 & -0.5 \end{bmatrix} = \ldots = 11$$

You may have noticed that the matrices in the examples above are all of the same size. That is not usually the case, with the filter matrix commonly being much smaller than the image matrix. In that case, the filter "travels" over the image, swiping it one step at a time. In the example in Fig. 4.16, we convolve a rectangular 4×3 image with a 2×2 filter:

And a numeric example:

$$\begin{bmatrix} 1 & 2 & 3 \\ 4 & 5 & 6 \\ 7 & 8 & 9 \\ 10 & 11 & 12 \end{bmatrix} * \begin{bmatrix} 1 & 2 \\ 0 & -1 \end{bmatrix} = \ldots = \begin{bmatrix} 0 & 2 \\ 6 & 8 \\ 12 & 14 \end{bmatrix}$$

The size of the final matrix will depend on two details. The first is the length of the step that the filter takes at each round. This is commonly known as **stride** and, in all examples above, was equal to one, meaning that the filter moved one cell to the

$$\begin{bmatrix} A & B & C \\ D & E & F \\ G & J & K \\ I & L & M \end{bmatrix} * \begin{bmatrix} \alpha & \beta \\ \gamma & \delta \end{bmatrix} = \begin{bmatrix} (A\alpha + B\beta + D\gamma + E\delta) & \\ & \\ & \end{bmatrix}$$

$$\begin{bmatrix} A & B & C \\ D & E & F \\ G & J & K \\ I & L & M \end{bmatrix} * \begin{bmatrix} \alpha & \beta \\ \gamma & \delta \end{bmatrix} = \begin{bmatrix} (A\alpha + B\beta + D\gamma + E\delta) & (B\alpha + C\beta + E\gamma + F\delta) \\ & \\ & \end{bmatrix}$$

$$\begin{bmatrix} A & B & C \\ D & E & F \\ G & J & K \\ I & L & M \end{bmatrix} * \begin{bmatrix} \alpha & \beta \\ \gamma & \delta \end{bmatrix} = \begin{bmatrix} (A\alpha + B\beta + D\gamma + E\delta) & (B\alpha + C\beta + E\gamma + F\delta) \\ (D\alpha + E\beta + G\gamma + J\delta) & \\ & \end{bmatrix}$$

$$\vdots$$

$$\begin{bmatrix} A & B & C \\ D & E & F \\ G & J & K \\ I & L & M \end{bmatrix} * \begin{bmatrix} \alpha & \beta \\ \gamma & \delta \end{bmatrix} = \begin{bmatrix} (A\alpha + B\beta + D\gamma + E\delta) & (B\alpha + C\beta + E\gamma + F\delta) \\ (D\alpha + E\beta + G\gamma + J\delta) & (E\alpha + F\beta + J\gamma + K\delta) \\ (G\alpha + J\beta + I\gamma + L\delta) & (J\alpha + K\beta + L\gamma + M\delta) \end{bmatrix}$$

Fig. 4.16 A step-by-step convolution between a 4×3 image and a 2×2 filter, resulting in a 3×2 matrix

right (or to the bottom) at each iteration. Longer steps imply fewer rounds, since the filter will not stop at each cell. As a consequence, the final matrix will be smaller.

A second configuration affecting the final image is how the convolution operation behaves when it encounters the borders of the original image. Three possible options are commonly available (Fig. 4.17):

- **valid padding**: this is the policy most commonly used, and is the same one implemented in the examples above. With valid padding as soon as the filter encounters the border of the image it stops and goes to the following row. The resulting matrix will be smaller than the original image.
- **same padding**: this policy works with reference to the center of the filter, which is the central cell if the filter is an odd-sized square (a very common case). It's the center of the filter that moves around inside the image, and thus a part of the filter can end up outside of the image. To perform the convolution the image is considered to have a value zero outside its border—in other terms, the image is zero-padded. The resulting matrix will be of the same size as the original image (if the stride is equal to one).
- **full padding**: with this policy, as long as there is at least one cell of the filter overlapping with the image, the convolution keeps being calculated. As in the previous case, the original image is zero-padded to accommodate for the out-

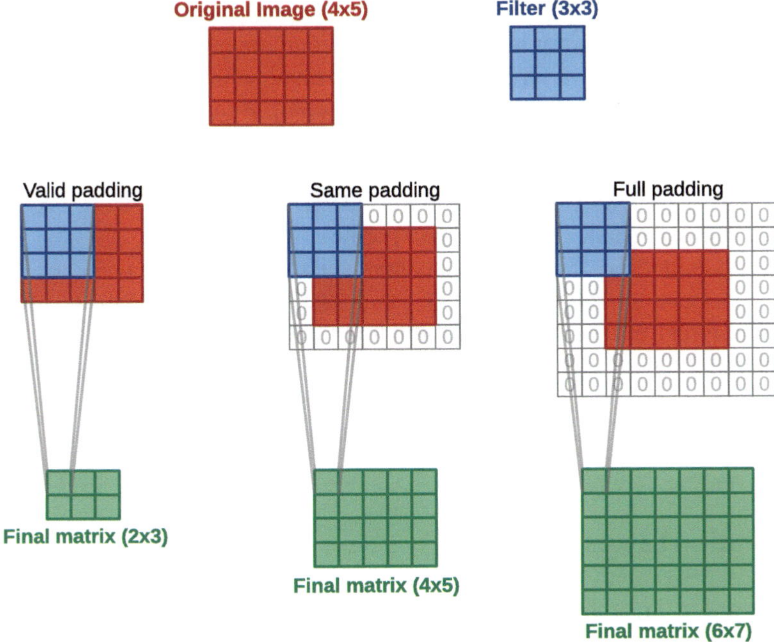

Fig. 4.17 The three available policies for padding. Here we assume that `stride` is equal to one

of-border pixels. The resulting matrix may end up being larger than the original image (it also depends on the `stride` value).

How a Computer "Sees" an Image

Computers manipulate numbers, and that's it. So, for a computer to work with images—from regular photos to X-rays to CT scans to multi-spectral images—it must encode the chromatic information in numbers.

The space is divided into a fine grid, each dot corresponding to a pixel. In the easiest case, a pixel contains only the information about light intensity, and when it is finally transmitted to a screen it will carry data going from completely black to completely white. A common convention is to use 8-bit encoding, and thus an image can be thought of as a big matrix containing values from 0 (full black) to 255 (full white). Each intermediate value will be some kind of gray, darker or lighter, depending on the numerical values contained in the matrix. We just defined a gray-scale image with 256 shades.

To obtain colors we note that visible light can be decomposed in simpler constituents. A very common format is RGB, with the letters standing for the three color channels Red, Green and Blue. An RGB image is basically three

(continued)

different images, one for each channel. By combining different intensities in the channels a wide variety of colors can be obtained. E.g. a pixel containing $R = 255,\ G = 0,\ B = 255$ will be brightly purple (namely: fuchsia), since by combining red and blue you obtain purple. If we take that pixel and halve all values to $R = 128,\ G = 0,\ B = 128$ we again obtain a purple, but darker (namely: patriarch). Do that once more and we obtain $R=64,\ G=0,\ B=64$, an even darker shade (namely: tyrian purple).

RGB is not the only color encoding available. You may encounter HSV (Hue, Saturation, Value), CMYK (Cyan, Magenta, Yellow, Black) and others. One different take is RGBA, with the first three channels containing regular RGB information and the fourth one (Alpha) describing how transparent the pixel should be.

Convolution Layers

A convolution layer is a network layer made up of several convolution units. Each unit performs a convolution between its internal filter and the incoming (multi-channel) data. When you convolve multi-channel (e.g. RGB) images, the convolutions shown in Fig. 4.16 are summed over layers (the unit will produce a single output matrix). Different units will contain different values in their internal filters, so each unit will produce a different result even if they all received the same input data. The output of the layer is formed by the output of all units being collated together, each unit representing a channel of the output data (Fig. 4.18).

A convolution layer *learns* by changing the values in the filters contained in each unit. The size of the filters is the same for all units, so it's easy to compute the total number of parameters. For example, a convolution layer with 5 units, each containing a 3×3 filter, will have 45 parameters to be tuned (not considering—for a moment—the bias terms, for the sake of simplicity). Notice that this is independent from the input data size and it's probably one of the most influential differences between convolution and dense layers: the number of parameters to be learned is reduced by several orders of magnitude! Table 4.2 reports some examples of how the number of parameters that need to be learned varies depending on input data and layer architecture (Fig. 4.18).

Each unit in a convolution layer receives the same input data as the other units, but once learning is completed each unit will be configured to look for different patterns in the image. A unit could be configured for, e.g., highlighting horizontal lines in the data. Another unit could look for vertical lines, and so forth. The output from the various units is then collated and forwarded to the next layer (Fig. 4.19).

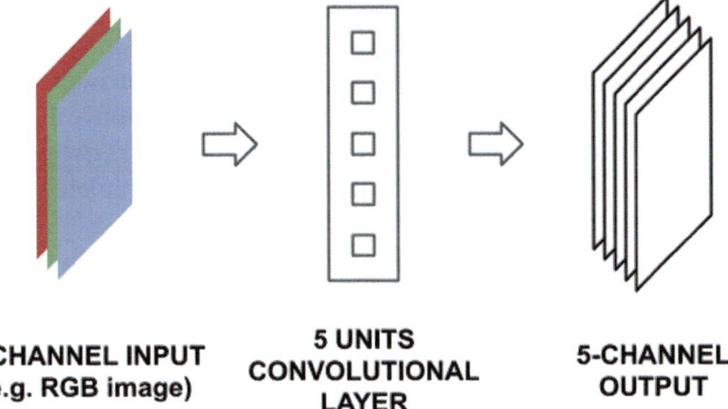

Fig. 4.18 The effect of a convolution layer on data size

Table 4.2 Comparing the number of parameters for two convolution layers and a dense layer (For the sake of simplicity, the bias terms are omitted: these, by the way, would be exactly the same in dense and CNN layers, one per unit)

	Number of parameters		
Input data	3×3 filter, 5-units convolution layer	5×5 filter, 10-units convolution layer	5-units dense layer
Small RGB image (50×50×3)	45	250	37,500
Medium RGB image (128×128×3)	45	250	245,760
Small multichannel image (50×50×5)	45	250	62,500
Generic data cube (20×20×20)	45	250	40,000

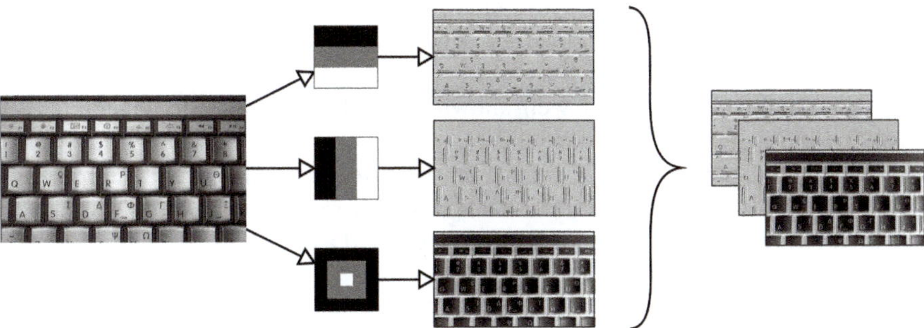

Fig. 4.19 Inside the same convolution layer different units learn to look for different patterns in the data. Each convolution node will produce a new channel, and all the channels will be fed to the next layer. Detail taken from public domain image "Greek Keyboard (Macbook Pro)" (Klilikakis https://commons.wikimedia.org/wiki/File:Greek_Keyboard_(Macbook_Pro).jpg)

Max Pooling Layers

A max pooling layer behaves in a fashion similar to a convolution layer, but the mathematical operation is simpler: it just takes the maximum value[9] contained in the filter. Then the filter is moved on to a new area, exactly as done for convolution layers. Thus, no learning happens: the layer output is completely defined by its hyperparameters.

A typical configuration for this layer uses "same" padding and a `stride` hyperparameter equal to the filter side. This implies that the resulting image is smaller. If the max pooling filter is 2×2, each side of the image will be halved. If it's 3×3 the output will be a third of the original size.

Flatten Layers

This is a very simple layer that again—just like pooling layer—does not learn anything. In fact, this is just wiring. Unlike the pooling layers, there aren't even any hyperparameters. Data coming out of convolutional layers will typically be in the shape of a three-dimensional cuboid. But the final, dense part of the network presented in Fig. 4.14 requires data to be in the shape of a one-dimensional array. A flatten layer does just that. Remember that cell order is not important for dense layers, since they cannot exploit data locality. So, a flatten layer just takes the cuboid-shaped data and slices it into a single big (long) noodle.

Batch Normalization Layers

This layer normalizes input across batches [83]. We haven't yet introduced the concept of a data batch, but let's just anticipate that, owing to memory constraints, it is often not possible to elaborate an entire dataset in a single pass. So data is processed in pieces—batches.

Data normalization is a very well-known concept in statistics. In general terms, normalizing means manipulating a data series so that it has known, expected features, e.g. a known average value, or the minimum value equal to zero. A common normalization takes a set of data points $x_1, x_2, \ldots, x_n$ and produces a new data series $\hat{x}_1, \hat{x}_2, \ldots, \hat{x}_n$, with each element defined as:

$$\hat{x}_i = \frac{(x_i - \mu)}{\sigma}$$

[9] In general terms there could be several types of pooling layers (e.g. *average*), but actually most of the time the *max* function is used. The intuition behind pooling is that large numbers encode important information on the input data, compared to zero (or almost zero) values (in this respect, *max* works better than *average*).

where μ is the mean and σ is the variance of the data series. In other words, the normalized series will have zero mean and unitary variance (data standardization).

The same normalization can be applied to an entire data batch coming from a previous network layer (data flowing between intermediate layers), so that when the data arrives at the next one it will have known characteristics. This usually has the effect of improving the learning process of a sequential network, both in terms of general performance and faster learning.[10]

Batch normalization helps to contrast outliers—either native to the input dataset or introduced by the network—and thus keeps the learning process stable. Other types of normalization can happen, e.g. a normalization layer could bring the data to a unitary L2 norm or some other statistical metric.

A Typical Architecture: Reprise

By this point, we described all the basic building blocks of a (sequential, convolutional) neural network and are ready to revisit Fig. 4.14. The data shape changes through the network. In particular, the padding strategies adopted in convolution and pooling layers influence the "face" of the data cuboid, typically shrinking it. On the other hand, the number of units in each layer commands the number of channels (i.e. the thickness of the cuboid). And since the number of units usually grows through the network, the cuboid becomes thicker and thicker.

So, why is this typically used as CNN architecture?

This is not a trivial question. Why do we focus on this type of architectures, sometimes called "funnel" networks for their effect on the data shape? The short answer is: because they work. There is also the side effect that this kind of architecture is easy to conceptualize and (somewhat) understand for a human network designer. But the core reason is completely heuristic: they work, well, and nobody knows exactly why they do, or why we should prefer this kind of network shape to something else. As a matter of fact, researchers try different networks all the time. ResNet networks, a family of networks responsible for a big improvement in the state of the art, are built on the network template we just discussed, with some extra connections.[11]

So, why, you may ask again?

This kind of network is very good at building semantic understanding. The only thing a network *can* actually do is to recognize patterns in the data. But the multi-

[10] In other types of network architectures—namely recurrent neural networks—batch normalization has detrimental effects. In that case, the network can benefit from *layer normalization*, which is similar in general terms but the normalization happens per single data point and not per batch [84].

[11] ResNet architectures incorporate identity shortcut connections (a.k.a. "skip connection") that connect the output of a layer directly to the input of a deeper layer, skipping a few layers in between. This mitigates the vanishing gradient problem and allows for much deeper networks to be trained effectively [85, 86].

layered architecture allows us to split a very complex problem ("is there a cat in this picture?") into simpler—and most importantly, hierarchical—steps.

More formally, semantic understanding in a neural network is typically achieved through the use of multiple layers, each of which is responsible for learning and representing different levels of abstraction in the data. The first layers of the network tend to recognize very simple patterns in the image, e.g. the presence of vertical, or horizontal, elements. Deeper layers learn more complex, abstract features such as object categories and relationships between objects.

4.7 Optimization

Playing chess well is an optimization problem, since at each round a player is called to select the best move among all possible ones. Optimization is exactly that: finding the best choice among all possible choices. In many real world problems (and in chess, too!), it's just plainly impossible to test all choices, so it is very important to have some kind of smart strategy that helps with the search.

In the field of neural networks, optimizing means finding the best set of parameters inside each layer of the network. This typically means adjusting the values of the weights and biases of the network's connections in order to minimize the error between the predicted and the true output of the network. For a gallery of possible error measures please see Sect. 3.1.

Whatever the selected error metric—more precisely, as you already know, whatever the selected *loss function*—there are many different optimization algorithms that can be used to train a neural network, each with its own set of advantages and disadvantages. Some of the most common optimization algorithms include stochastic gradient descent (SGD), ADAM, and RMSprop. These algorithms use techniques such as gradient descent, momentum, and adaptive learning rates to adjust the network's parameters in order to minimize the error. The specific mathematics can become a little intense, and the specific implementation can become tricky very quickly. Let's just visit a few key concepts.

The Limits of Gradient Descent

In Sect. 4.3, we introduced the gradient descent algorithm in its most basic form. From the practical point of view, nobody really uses gradient descent as it is, because of two main limitations:

- **it's slow**: you may recall that the length of the step taken by gradient descent is proportional to the local derivative (i.e. the gradient). Since the derivative shrinks the closer you get to a minimum, the algorithm ends up stuck in an "Achilles-and-the-tortoise" situation: the closer it gets to the target, the slower it becomes (see Fig. 4.20)

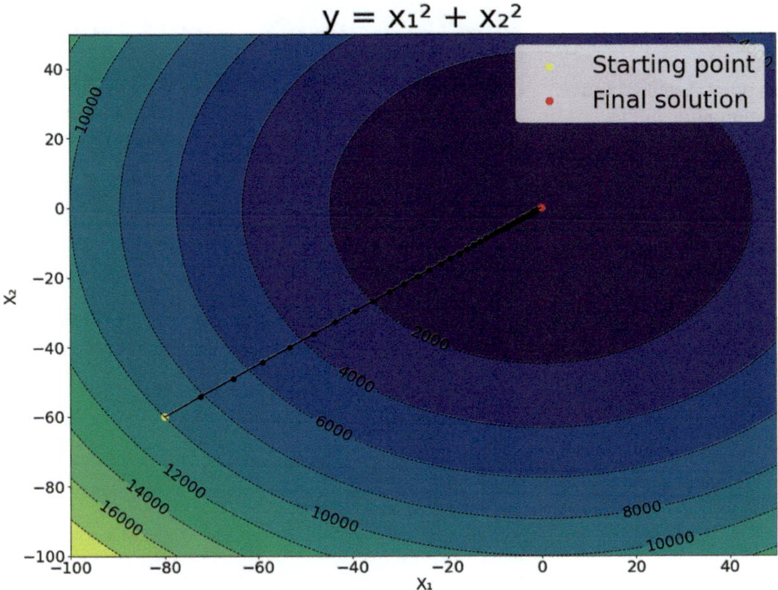

Fig. 4.20 The trajectory of the gradient descent algorithm. Notice that the length of the step becomes smaller and smaller the closer the algorithm gets to the target

- **it's easily tricked by local minima**: there's no way for the algorithm to overcome even the smallest ridge separating two minima. In fact, once it enters "the orbit" of a local minimum, the algorithm's behaviour is completely defined, and it will stop only when it reaches the bottom of the valley. Moreover, the algorithm is heavily influenced by the starting point, and thus starting the exploration from different places can have drastic consequences for the final result (Fig. 4.21)

A number of algorithms have been proposed to overcome the naive gradient descent limitations. We describe two of them: RMSprop and ADAM.

What Is Exponential Average?

Imagine you receive temperature data from a distant city. Every day you get a number, in degrees, and you are required to compute an average temperature for the location that takes into account the historical data and provides an estimation of, well, the average temperatures.

You could do a grand mean, adding all the historical data and then dividing for the number of days. This plain average certainly works, but with two limitations: first, cycles in the data (e.g. seasonal fluctuations) are averaged

(continued)

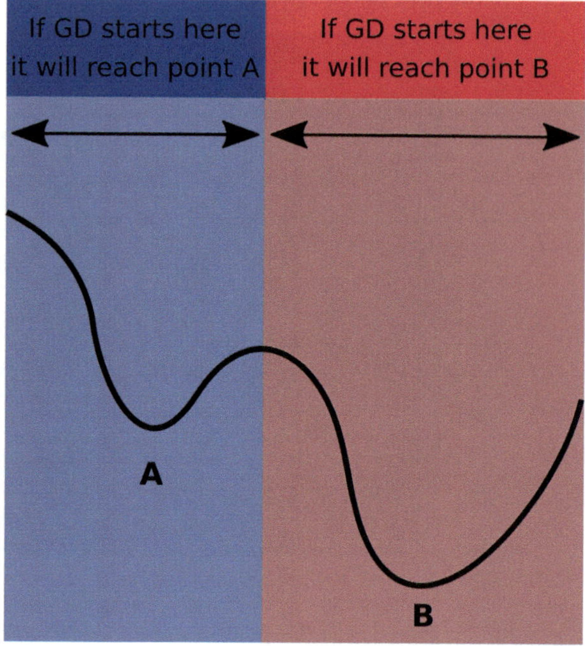

Fig. 4.21 Gradient descent is completely incapable of overcoming the ridge between point A and B, and its final minimum will be determined by where the exploration starts

out, since very hot summer weather and very cold winter weather just cancel each other out. Second, if you have many historical data it will be very slow to pick up trends, e.g. the rising of temperatures due to climate change.

So, you may want to use a different approach, one where the last few days or weeks have a greater importance. Instead of averaging the new daily temperature with *the entirety* of historical data a simple solution could be to keep track of just the last few data points, e.g. the last week, or month, or season. This approach is called a sliding window average (or rolling average) and allows us to pick up trends in the data depending on the size of the window. It has the small technical disadvantage that it requires keeping track of the last data points.

An even simpler approach is to use the exponential average (EA). First, you fix a decay parameter $d \in [0, 1]$ that describes how quickly (or slowly) your average will respond to changes, with 0 being extremely slowly and 1 being extremely quickly. Then, for each new incoming data point, you compute a weighted average between the exponential average calculated so far and the new data point:

$$\begin{cases} \mathrm{EA}_1 = t_1 \\ \mathrm{EA}_i = d\mathrm{EA}_{i-1} + (1-d)t_i \end{cases}$$

(continued)

where t_i is the temperature for day i and EA_i is the exponential average computed for that day.

This method has the advantage of a very simple implementation in which only one value must be stored (the last average), with the additional benefit that the parameter d has a very intuitive interpretation.

RMSprop

RMSprop (Root Mean Square Propagation) improves the gradient descent algorithm by adapting the learning rates based on the exponential average of the recent gradients (squared). This can help to prevent oscillations in the optimization process and allows the model to converge more quickly to a good solution.

To implement RMSprop, you will need to maintain a running average of the squares of the gradients. At each training step, you will compute the gradient of the loss function, and update the exponential average. You then use this running average to normalize the learning rate and update the model parameters. More formally, at iteration i the algorithm will be testing a configuration of parameters W_i via a cost function $J(W_i)$ with a gradient $\nabla J(W_i)$. You then compute:

$$\begin{cases} \text{EASG}_i := d\text{EASG}_{i-1} + (1-d)\nabla^2 J(W_i) \\ \Delta_i := -\frac{\lambda}{\text{EASG}_i}\nabla J(W_i) \\ W_{i+i} = W_i + \Delta_i \end{cases}$$

where EASG_i is the exponential average of the squared gradient for iteration i, λ is the learning rate, Δ_i is the updating step that the algorithm will take at iteration i, and d is the decay parameter for the exponential average.[12]

Note that the only difference with regular gradient descent is that the learning rate λ is scaled by the value of EASG. In fact, if you set $d = 0$ you fall back exactly to the gradient descent algorithm. The rescaling of λ implies that when the gradient becomes small the algorithm doesn't get completely stuck, as happens with gradient descent, since a small $\nabla J(W_i)$ is compensated by a large λ-to-EASG ratio.

ADAM

The ADAM (ADAptive Moment estimation) method is similar to RMSprop but adds a second exponential average when computing the updating step Δ_i. The learning rate λ is scaled exactly as in RMSprop, but ADAM doesn't directly use the gradient.

[12] Depending on the software library you use, the parameter d is sometimes called *rho*.

Instead, it uses its exponential average. More formally:

$$\begin{cases} \text{EASG}_i := d_1\text{EASG}_{i-1} + (1 - d_1)\nabla^2 J(W_i) \\ \text{EAG}_i := d_2\text{EAG}_{i-1} + (1 - d_2)\nabla J(W_i) \\ \Delta_i := -\frac{\lambda}{\text{EASG}_i}\text{EAG}_i \\ W_{i+i} = W_i + \Delta_i \end{cases}$$

where EAG_i is the exponential average of gradients for iteration i. Note that, since we are computing two different exponential averages (EAG and EAGS), we need two decay parameters d_1 and d_2.[13] Intuitively, when gradients are shrinking (i.e. moving towards flatter areas) EAG increases more quickly than EASG (square of small gradients): therefore, the EAG/EASG ratio increases and gives a boost to the updating value Δ, thereby allowing the optimization to move further and faster. On the contrary, when the gradients expand (i.e. moving on steeper slopes), the opposite happens: the EAG/EASG ratio decreases and keeps in check the updating value Δ, thereby preventing the optimization from going too fast and missing the minimum (overshooting). ADAM operates a bit like the accelerator and the brakes of a vehicle, giving a finer control to the optimization process.

The ADAM algorithm builds on the results of RMSprop and adds robustness to problems with sparse gradients, i.e. flat areas with very small gradients, which can be challenging for RMSprop and a disaster for regular gradient descent.

The Role of the Learning Rate

The learning rate is a crucial hyperparameter: it determines the step size at each iteration when updating the model weights, and ultimately controls how much the model changes in response to the estimated error during training.

A proper learning rate ensures that the model converges to a minimum in a reasonable time. If it's too high, the model may fail to converge, or may end up oscillating around a suboptimal solution. Conversely, if the learning rate is too small the convergence process will be slow, requiring many iterations and extensive computational resources.

Figure 4.22 shows the effect of the learning rate on a simple two-dimensional polynomial problem. The top left panel shows a good choice of learning rate: the algorithm follows the feature landscape towards the minimum and, once this has been reached, it stays there. The top right panel shows a learning rate that is slightly too high: the algorithm is able to arrive close to the desired solution but then keeps oscillating. In the bottom left panel the learning rate is too small and, even if the algorithm is on the right track, it ends up consuming all the available epochs before

[13] Depending on the software library you use, the parameters d_1 and d_2 are sometimes called β_1 and β_2.

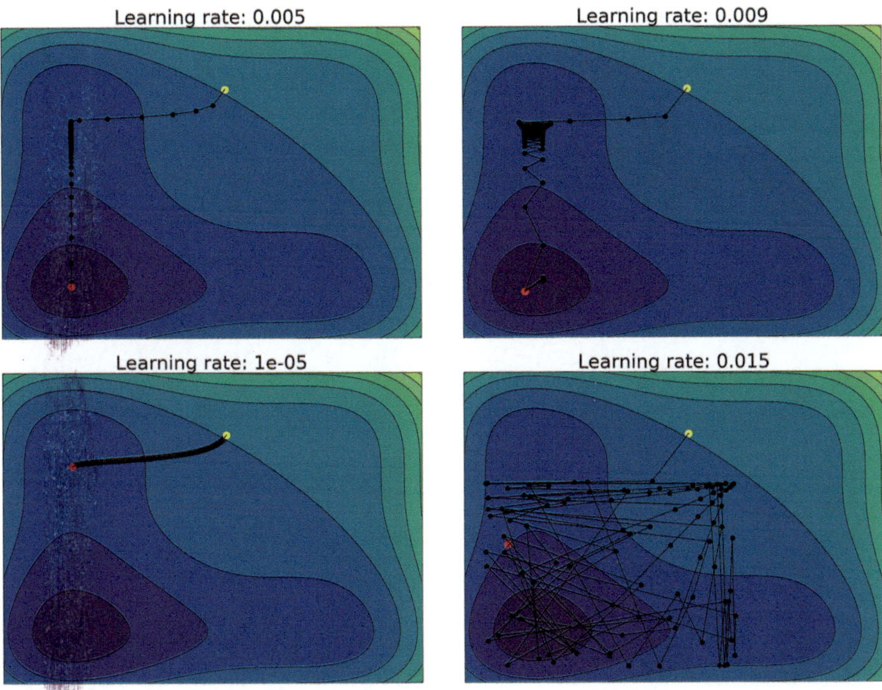

Fig. 4.22 The effect of the learning rate. In this example naive gradient descent is applied to minimize a two dimensional polynomial cost function (darker colors represent better results). For all panels, the algorithm starts from the yellow dot and terminates with the red dot

getting close to the target. In the bottom right panel the learning rate is just too high and the algorithm jumps all over the place.

Unfortunately, we are almost never so lucky as to be able to plot the feature space. We thus must determine whether the choice of learning rate is adequate through indirect observation. Figure 4.23 shows the effect of the learning rate on the most commonly monitored observable: the loss. The top left panel shows a learning rate that is clearly too high: the loss oscillates a lot, both in the training and test sets, and there's no clear sign of convergence. The top right panel shows a better choice for learning rate. Most of the actual learning happens in the first few epochs, then the system stabilizes around a solution and only marginal improvements find place. The few spikes still present may indicate that the learning rate is still a bit too high, though. The bottom left panel shows a solid choice of learning rate, no spikes present. However, examining the absolute values of the loss function we see that the final values are higher, in absolute terms, compared to the previous attempt. This solution is still improving when it reaches the limit of 200 epochs (declining slope). The bottom right panel shows a learning rate that is too small, and very little learning happened in the allotted amount of epochs.

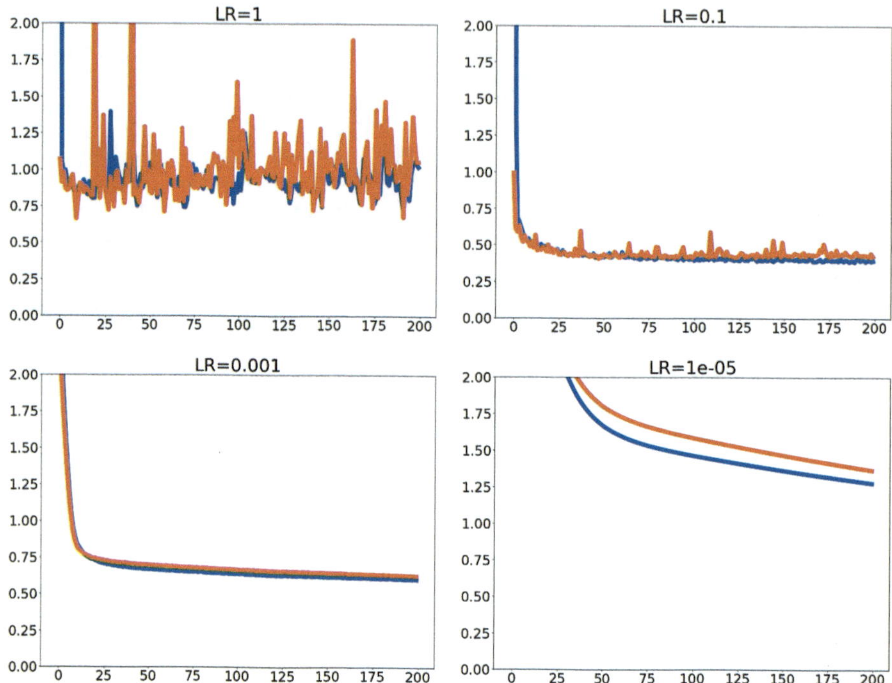

Fig. 4.23 How Learning Rate (LR) influences the loss function in a classification problem. On the X axis there are the training epochs; the line color represents the loss for the training (blue) and validation (orange) sets

A hidden difficulty of high dimensional feature spaces is that they can become so complex as to make it simply impossible to find the "right" learning rate, i.e. a value that balances between the speed of convergence and the accuracy of the model. Together with adaptive optimizers (like Adam and RMSprop), useful tools for optimising the model in complex scenarios are the so-called Learning Rate Schedulers. In the typical use case, a scheduler gradually decreases the learning rate over time so as to fine-tune the weights as the model converges. The intuition behind this approach is that a higher learning rate helps with a general exploration of the feature space, while smaller values help find the actual local minimum, where a more thorough search of the surrounding space is likely to be beneficial.

4.8 Overfitting and How to Contrast It

We introduced overfitting in Sect. 3.2, when we discussed the importance of cross-validation and data resampling to avoid the "overfitting trap". Let's say that we want to predict the probability that a patient will die from or survive a disease based on some clinical and lifestyle data. In this hypothetical example, we have historical (training) data on 1000 patients, and we observe that 900 of them survived the disease while 100 of them unfortunately died. This translates into an overall survival rate of 90%. This is very simple (and we are probably not overfitting the data), but is also obviously biased: patients can be men or women, old or young, they can be overweight, have the habit of smoking, engage in more or less physical exercise, suffer from concurrent comorbidities etc. Therefore we refine our model and look at old men (>70 years of age): there are 300 such patients, 100 died and 200 survived the disease. This means that our probability of survival went from 90% to 66.6%, which is less biased and probably more precise and useful. Let's now take a big step forward and use many more variables to make a much less biased comparison: we focus on old men (over 70) that are overweight, who smoke, don't exercise, don't drink alcohol, with high socio-economic status, who are independent professionals, who live in a residential neighbourhood, with one known chronic comorbidity etc. Using this virtually unbiased and very precise set of variables, we are left with only three patients, who all died: the probability of survival is now 0%, but would we feel comfortable with this result? And would we assign a 100% probability of dying to a new patient with the same characteristics as the three patients in the training set? Obviously not, and the reason is that our model is perfectly describing the training data (zero patients with those characteristics survived the disease), but has no use with new data. The bias was reduced, but the cost was a dramatic increase in the variability, as an estimate based on only three observations has large variance ($\frac{\sigma^2}{\sqrt{N}}$): if we take a different training set of patients with the same characteristics, the estimated probability of survival will almost certainly be very different from zero.

We said that overfitting is prone to occur if: (i) we use the **same data to fit (train) the model and to make predictions**; (ii) we **overparameterize the model**, e.g. if we include too many variables (like the millions of features from 'omics data); and (iii) we use **flexible methods** (e.g. polynomial functions, splines, classification trees etc.) that yield highly complex models.

As we already mentioned a few times earlier in this book, neural networks are outstandingly powerful at approximating very complex functions and capturing non-linear relationships and patterns in the data. As a result, neural network models naturally tend to be very sophisticated and complex and, also naturally, are very prone to overfit the data. Therefore it is of vital importance to control overfitting: this is done by employing cross-validation and also by resorting to additional techniques that are known collectively under the name of **regularization** (a.k.a. penalization). The idea is to "penalise" the model on the training data by making its task (understanding the relationships in the data) more difficult. We describe below a few such methods, although several more exist.[14]

[14] Mental note: data augmentation can also be thought of as a way to control overfitting. Other approaches to control overfitting do exist, for example the addition of noise to the data (noise regularization [87]): we restrict ourselves here to the most common methods.

Regularization: L1/L2

L1 and L2 regularization are techniques commonly used in statistics to prevent data overfitting. The core idea is to penalize the adoption of large values for model parameters, e.g. the coefficients of a dense neural network model.

In neural networks L1 and L2 regularization are added to layers that actually learn something, i.e. dense and convolution layers. It would not make sense to add regularization to purely predetermined layers like pooling or flattening layers.

L1 regularization, also known as Lasso regularization, adds a penalty term to the loss function that is proportional to the absolute value of the model's weights. It is defined as:

$$\mathrm{L1}_{penalty} = \lambda_1 \sum_i |w_i|$$

where λ_1 is the regularization strength hyperparameter and w_i are the weights of the model.[15]

L2 regularization, also known as Ridge regularization, adds a penalty term to the loss function that is proportional to the square of the model's weights. It is defined as:

$$\mathrm{L2}_{penalty} = \lambda_2 \sum_i w_i^2$$

where λ_2 is the regularization strength hyperparameter.

During backpropagation the two regularization terms are simply added to the loss function, making the minimisation job harder for gradient descent (and derived algorithms):[16]

$$\text{Penalized Loss} = \text{Loss} + \mathrm{L1}_{penalty} + \mathrm{L2}_{penalty}$$

Both L1 and L2 regularization are useful for preventing overfitting, and which one to use depends on the specific problem and the desired properties of the model. L1 regularization is generally more effective at reducing overfitting when the number of features is large, while L2 regularization generally works better when the number of features is small.

[15] Attention! Don't mix up the regularization parameter λ_1 (and later λ_2) with the learning rate λ. Unfortunately this is the most common notation, and thus we had to stick to it. Just keep in mind the difference between optimization and regularization and you'll be good.

[16] You can add either one of, or both, L1 and L2 penalisations to the loss function.

Regularization: Dropout

Dropout is a regularization technique for reducing overfitting in neural networks that was introduced by Srivastava et al. in 2014 [88].

In dropout regularization, neurons (units) in the neural network are randomly "dropped out" or "turned off" during training with a certain probability, typically between 0.2 and 0.5. It works because it forces the network to rely on a reduced number of units and forces the units that are not dropped out to learn more robust features. This has the effect of reducing the complexity of the model and preventing overfitting.

During testing all units are used. During training the output of the units is scaled up by the dropout rate, so that the overall sum is unchanged.

Regularization: Early Stopping

Early stopping is a somewhat problematic technique, since it (partially) violates the separation between training and validation sets. That said, it is often useful and often used, so we decided to include it in our showcase of regularization techniques.

It starts from the observation that loss values often show the pattern illustrated in Fig. 4.24.

This pattern indicates that for the first few epochs the model is learning "well", in the sense that both the training loss and the validation loss shrink. After a certain point, however, the two curves separate: the training loss keeps going down while the validation loss starts growing. This means that the network has stopped

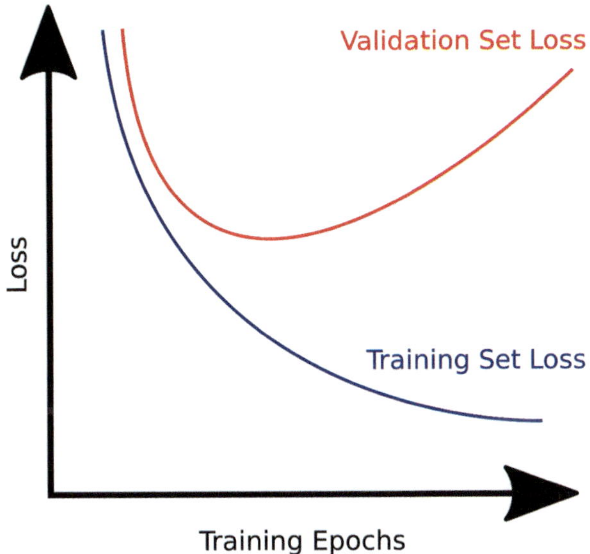

Fig. 4.24 A common pattern seen in the evolution of loss computed on training and validation sets through the training epochs. The loss in the training set decreases monotonically, while the loss in the validation sets reaches a minimum, then picks up again and starts increasing

generalizing and started overfitting. In other words, the network is learning the training set samples, not the underlying problem structure. This pattern in the loss function is very common when the network is too big for the combination of problem and sample size. One option is to redesign the network to be optimally-sized. Another option is to just stop when the validation loss starts to rise. That's exactly what early stopping does.

The exact implementation details will depend on the deep learning library. A common approach is to define an integer hyperparameter called *patience* (usually in the 5–10 range) and stop the training when the validation loss doesn't improve for that many epochs. This prevents a single bad training epoch from triggering early stopping.

Ideally the trained model should then be tested on a completely new batch of samples (the "test set") since, formally, the validation set has been involved in the training process.

Recurrent Neural Networks (RNN) 5

We have so far discussed in great detail two of the most common types of neural network architectures: those based on dense layers (fully connected neural networks) and those based on convolutional layers (CNN: convolutional neural networks). We now introduce **recurrent neural networks** (**RNN**), which are yet another common type of architecture in deep learning.

RNN models have been developed to work with **sequence data**: this is a type of structured data where order matters, i.e. it is relevant whether a piece of information is placed before or after another piece of information (see the following box "Structured or unstructured?" for a definition of structured data). Examples typically include text data (sentences in an article or book are read from left to right and from top to bottom—at least with the Latin and Cyrillic alphabets) and sounds (a song is an ordered succession of musical notes). We can also find clear examples of sequence data in molecular biology, like the sequence of nucleic acids (DNA, RNA) or of proteins, or in time series data like recorded temperatures and rainfall (climatology), the trend of the incidence of diseases over time (epidemiology), heart rates from electrocardiography (medicine), exchange rates between currencies (econometrics). Also, in agriculture, the emerging field of smart farming/precision agriculture is based on technological devices and sensors that produce streaming data with a longitudinal component (i.e. a sequence of data): examples include cow milk-yield measured continuously from milking robots or AMS (automated milk-recording systems) and the automatic recording of animal movements or rumination time with sensors.

In this chapter, we are going to discuss the basic RNN architectures and models for the analysis of sequence data, and some specialised types of units and layers that have been developed specifically to address challenges related to the longitudinal aspect of sequences.

F. Biscarini, N. Nazzicari, *Deep Learning for Life Sciences*, Decoding Evolution 1,
https://doi.org/10.1007/978-3-031-96852-5_5

5.1 Sequence Data Problems and Representation

With respect to the type of input data and output results, in sequence problems we can identify the following possible basic scenarios:

- input sequence $\rightarrow$ output sequence
- input number/category/string $\rightarrow$ output sequence
- input sequence $\rightarrow$ output number/category/string

For example, in **speech recognition** we have an input audio sequence (voice) and produce as output the corresponding text. On the other hand, in **music generation** we have an input string or category (e.g. the type of music we want to generate, like 'jazz' or 'rock') and produce an output audio sequence. Finally, in **text classification** we have an input text sequence (e.g. a newspaper article) and as output the corresponding category (what is the article about? Sport, cinema, politics, economy etc.).

As you may have noticed, these are all examples of supervised learning problems: at training, we have both the input data (e.g. voice recordings, songs, newspaper articles) and the answers (e.g. voice transcripts, corresponding labels for songs and articles).

Now, as usual in deep learning (although this holds for data analysis in general) we have the problem of data representation: how do we encode sequences so that the neural network model (and ultimately the computer) can understand them? Let's take the example of text (sentences) as illustration: one possibility is to use a (large) dictionary of words in a language and then represent each word in a sentence as a (long) one-hot encoded (OHE) vector of 0's and 1's: Fig. 5.1 illustrates the OHE-representation of an example input sentence, with a tiny vocabulary.

The input sequence will then be mapped to the corresponding output y (e.g. the label "this is a `FALSE` statement": statisticians are actually despicable human beings). Naturally, there is no rule on using words instead of, for instance, the

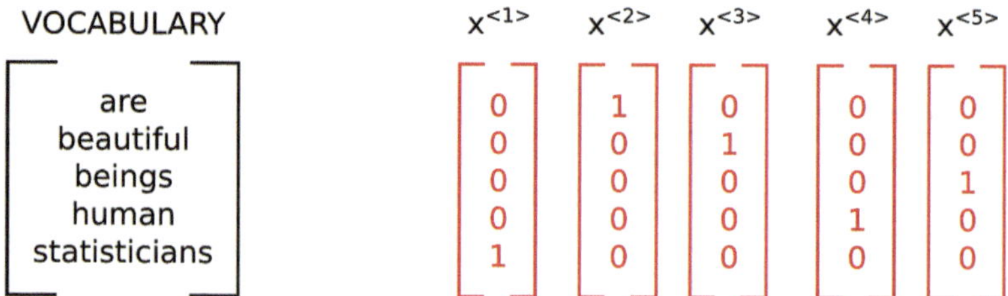

Fig. 5.1 Illustration of the one-hot encoding (OHE) representation of an input sentence x. The vocabulary lists the available words, e.g. in alphabetical order, which are mapped to the words in the input sentence

single letters as input data: this would be a perfectly valid alternative representation with its own pros (smaller 'dictionary' of letters) and cons (much longer sequences to model). Other more advanced representations of sequence data do exist, like embeddings and autoencoders, and will be briefly mentioned in later chapters (actually, that of data representation for input data in neural network models is an intense field of scientific research).

Structured or Unstructured?

Neural networks, be it CNN, RNN or other types of networks, are often said to work very well with "unstructured" data, typically with reference to natural language (articles, books), sounds (voice, music), images, videos. The reason is that these data would not fit easily in the structure of a relational database (RDB) and are not neatly organised in rows and columns like tabular data, with multiple tables possibly linked through key-variables. But are these types of data really unstructured?

We believe that calling these types of data "unstructured" is misleading, as they are actually rather the opposite: text, sounds, images, videos etc. have a strong inherent data structure since the order in which words in a sentence, notes in a song, pixels in an image or frames in a video occur, does matter. Nearby elements are related, and preserving this information in the model is highly relevant. This is unlike tabular data or relational databases, where the order of columns (or fields) can be swapped with no consequence.

Therefore, we propose to call tabular and RDB data **formatted** or **organised data**, and reserve the term **structured data** for language (text, voice), sounds, images, video etc.

5.2 Neural Networks for Sequence Data

Once we have a working data representation, the next question we face is how to design the **neural network model** for sequence data. We can start by trying the standard fully connected (dense) neural network, as in Fig. 5.2: the different words in the sentence are the x input variables that enter each unit in the first hidden layer, to then be processed by a variable number of other dense layers until the final softmax layer classifies each word to the corresponding label.

This model will work, operationally, but a number of problems would emerge. First, inputs and outputs may have different lengths in different sentences: this could be circumvented with zero-padding (adding 0's) to bring every input to the size of the longest sentence, but it is evident how this solution would be largely sub-optimal. Additionally, the number of learning parameters would quickly soar: if we take for example n short sentences (the size of the input data won't affect the number of parameters to be learnt) with a maximum length of 13 words (to just

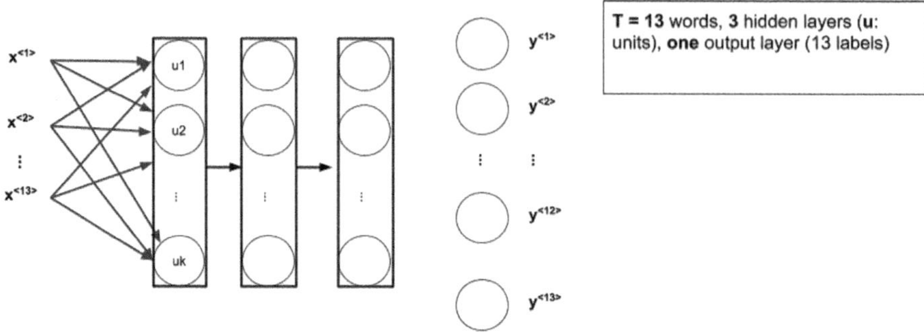

Fig. 5.2 A dense—fully connected—neural network for sequence data. The single words in the 13-word-long sentence are the input x variables, which are fed to each unit in the first hidden layer. A number of successive dense layers may follow, until each input word is classified to the corresponding output label

pick a number) and a relatively small vocabulary of 10,000 words, a NN model for binary classification (e.g. `TRUE/FALSE` statements) with one 16-unit hidden layer, the total number of trainable parameters will already be 2,080,033. This is because the input variables will be the 10,000 words in the vocabulary times the maximum length of the sentence, 13: $10{,}000 \cdot 13 = 130{,}000$. This large number of variables will enter each of the 16 units, yielding $130{,}000 \cdot 16 = 2{,}080{,}000$ weights plus 16 bias terms. Finally, the binary output layer will have one unit that receives the 16 'new' variables plus one additional bias term (vocabulary size × maximum sentence length × number of units × number of layers: see Remark 5.2.1 for a comparison of trainable parameters in RNN vs dense NN).

Most importantly, a dense neural network would not model the order in the sentence, thereby not transferring the learning along the sequence: i.e., the dense neural network will not be able to understand that a verb follows the subject and, more generally, that words in the sentence are logically ordered.

Similar to what we saw with convolutions and CNN (Section "A typical (?) CNN (Convolutional Neural Network")), we would like (and need) learning to generalize from one part of the sequence (image) to other parts of the sequence (image). This brings us to **recurrent neural networks** (**RNN**). Figure 5.3 demonstrates the idea behind RNNs: the elements of the sequence (e.g. words) enter the network sequentially, one at a time. Each element $x^{<t>}$ is fed to the RNN together with the activations from the previous 'layer', $a^{<t-1>}$, and these two pieces of information (current data + previous activations) are used by the model to predict $y^{<t>}$, the current target output. This is known as "recurrence" and is a key aspect of RNNs: the use of previous activation values together with the current data is the way by which RNNs **model the order of the elements** in the sequence, thereby **transferring learning along the sequence** and "memorising" what they "saw" in the earlier bits of the sequence. As you have certainly gathered, the superscript <t> refers to the index of the element in the sequence: $x^{<1>}$ (or $a^{<1>}$, $y^{<1>}$) is the first element

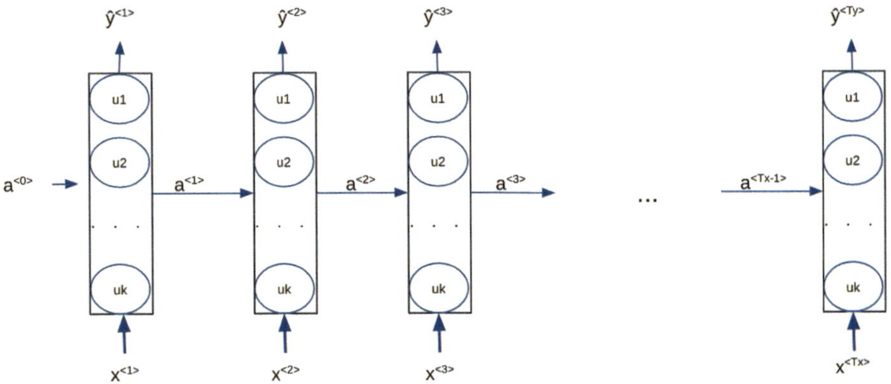

Fig. 5.3 A recurrent neural network (RNN): words (elements of the sequence) enter the network sequentially, i.e. one at a time. Each word $x^{<t>}$ is accompanied by the activation values from the previous 'layer', $a^{<t-1>}$ to predict the target $y^{<t>}$

(e.g. word) in the input sequence (or the first set of activations, or the first predicted output), $x^{<2>}$ is the second element in the sequence, and so on.

To be frank, Fig. 5.3 is misleading: it is useful for conveying the fundamental idea behind RNN, that of sequentially processing the elements of the sequence, thereby maintaining (and using!) the inherent order in the data; however, one might get the wrong impression that each element is processed by a different layer in the network, which is not so. A more realistic representation of what really happens in a RNN layer is given in Fig. 5.4: a single RNN layer, with u units, takes as input the elements of the sequence one at a time (sequentially), reproducing the order that is present in the original data sequence. Together with each element of the sequence, the activation values from the previous iteration are used to make predictions at time (position) <t>, thereby transferring learning along the sequence and keeping track of what has already been seen by the network ('memory'). Figure 5.4 also provides a better graphical visualization of "recurrence".

It is easily seen that, unlike dense NN, with a RNN it doesn't matter if input sequences have different lengths: elements are processed sequentially, and if a sequence happens to be longer/shorter this only means that there will be more/fewer iterations of calculations in the network, without any need for awkward zero-padding or constraints on the input sequences to all be of the same size. This also gives flexibility with respect to the output sequence, which can be of a different size than the input sequence: if you are wondering why we would want to have different input and output lengths, think of deep learning models for translation from one language to another, in which the number of words used in a sentence can be very different.

You can see that in Fig. 5.4 we have placed the weights (parameters) of the model: **Waa** for the activation values, **Wax** for the input data, **Way** for the target prediction. An important aspect to notice is that the very same sets of parameters are shared

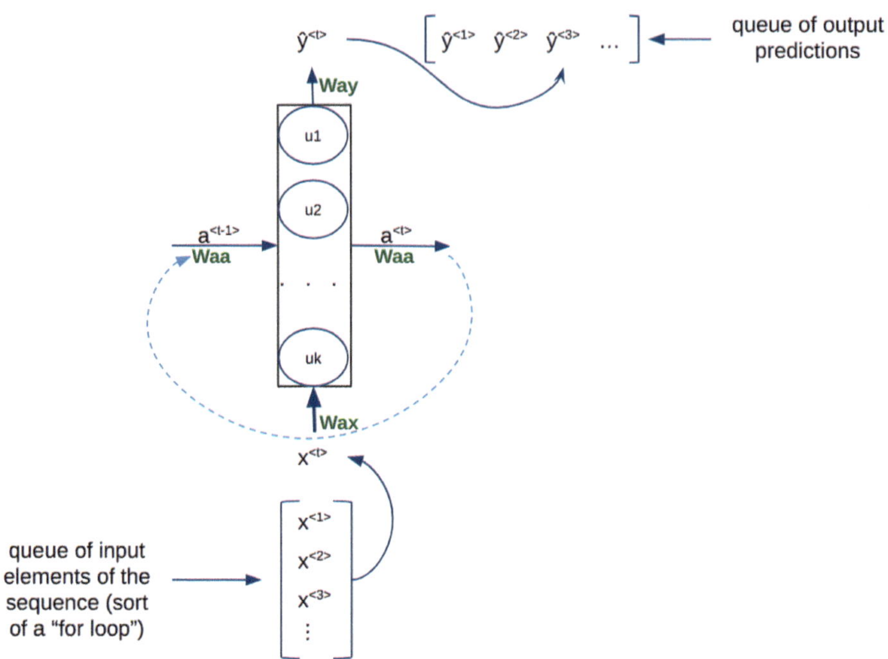

Fig. 5.4 The 'truth' about RNN layers: a single RNN layer sequentially takes as input the elements of the sequence ($x^{<t>}$), together with the activation values from the previous iteration ($a^{<t-1>}$), to predict the current target output $y^{<t>}$. **Waa**, **Wax** and **Way** are different sets of weights (model parameters) shared along the sequence. The dashed line shows that current activation values are used in the next iteration together with the next data element (recurrence)

across the sequence of data (i.e., are common to all elements of the sequence): this is what transfers the learning along the sequence and explains why we can reduce the number of trainable parameters (compared to a dense NN) and why we don't have problems with uneven sequences (sequences of different length).

Equation 5.1 reports the calculations involved in a RNN layer (for one input sequence: $n = 1$):

$$\begin{cases} a^{<t>}_{(u,1)} = g\left(\mathbf{W_{aa(u,u)}} \cdot a^{<t-1>}_{(u,1)} + W_{ax(u,m)} \cdot x^{<t>}_{m,1} + b_{a(u,1)}\right) \\ \hat{y}^{<t>}_{(1,1)} = \sigma\left(W_{ay(1,u)} \cdot a^{<t>}_{(u,1)} + b_{y(1,1)}\right) \end{cases} \tag{5.1}$$

Let's check what Eq. 5.1 tells us: for the moment, we ignore the terms in red. We see that we are left with an expression (in black) that is very similar to what we already saw in Eq. 4.2 for a basic single unit, or in Figs. 4.12 and 4.13 for the case of the fully-connected (dense) neural networks: we first multiply the input data by some weights (specific to each feature and unit) and add unit-specific bias terms, and the results are then transformed by the chosen activation function. In layer 2 (one unit), we then use these activations as 'new input data' that are multiplied by another set of weights, summed with a specific bias term, and activated to give us the prediction (the target at position <t> in the sequence). In a RNN, though, there is one important twist relative to the base dense-layer calculations: the terms in red are the **activation values from the previous layer/iteration** (<t-1>), which enter the calculations of the activations at position <t> with a new set of specific coefficients with size (u, u) (a square matrix of size u^2, with u = number of units in the layer). The coefficients $W_{aa(u,u)}$ are known as **recurrent weights** and are the key way by which RNNs are able to keep track of the order in the sequence (they sort of "remember" what was seen in the previous positions of the analysed sequence). As for the activation functions used in RNN, in the intermediate layers common choices are *ReLU* or *tanh* (**hyperbolic tangent function**: see Section "Activation functions" for details), while in the output layer the *sigmoid* or *softmax* functions can be used, depending on the type of output to be predicted (or generated).

Remark 5.2.1 At the beginning of Sect. 5.2, we saw that if we used a dense neural network to analyse sequence data—besides ignoring the order in the sequence—the number of trainable parameters would rapidly increase up to the point of becoming unmanageable. In the example of sequences with a maximum length of 13 elements, and a smallish vocabulary with just 10,000 words, a dense neural network with 16 units in the first layer and one unit in the output layer will have to learn 2,080,033 parameters. When we switch to a RNN, we first lose the maximum length of the sequence, since elements enter the RNN one at a time, in an ordered succession: we get rid of this stricture and use a more flexible architecture that processes sequences irrespective of their (possibly differing) lengths. The number of features then reduces to the size of the vocabulary: 10,000 words. From Eq. 5.1, we then see that W_{ax} is $10{,}000 \times 16$ units $= 160{,}000$ parameters. The number of **recurrent weights** W_{aa} is $16^2 = 256$. Finally, adding the 16 bias terms we have 160,272 trainable parameters in the intermediate layer. The output layer remains unchanged (16 'new' variables—the output of the 16 units in the intermediate layer—plus one additional bias term). This gives us a total of 160,289 parameters to be learned. This is 1,919,744 parameters fewer compared to the dense neural network implementation, a 92% reduction!

5.3 Backpropagation Through Time

In the Section "Forward and back propagation", we introduced backpropagation, the algorithm that updates the weights of a (dense) neural network so that the results (predictions) are incrementally corrected based on the known outputs and an error metric (remember: this is supervised learning). **Backpropagation through time (BPTT)** is a similar algorithm applied to RNNs and sequence data, and hinges on the same conceptual steps: (i) run the data through the network (*forward propagation*) to produce an output (predictions); (ii) measure the prediction error against the known outputs (this is done by using an appropriate **loss function**); (iii) calculate the partial derivatives of the loss function with respect to the network weights (going backwards through the network: *backpropagation*); (iv) update the weights to reduce the error; (v) repeat (*epochs*).

Compared to backpropagation for dense NN, with RNN we also need to consider the flowing of gradients over time (discretized into the t elements of the sequences): basically, we have one additional dimension to consider in our tensor calculations (and unfortunately, the more dimensions there are, the harder it is for the human mind to conceptualize and mentally visualize the problem). Anyway, forward propagation works as mathematically described in Eq. 5.1, for a single sequence (time) output $\hat{y}^{<t>}$. For backpropagation, we first need a loss function $L()$ to evaluate the model's performance: any one would do for this hypothetical illustration (specific loss functions will be chosen depending on the type of problem: probabilities, classification, regression, content generation etc.). We now start moving backwards in the network: from the loss to the prediction, from the prediction to the hidden state (activations) at position <t>, then from the tth hidden state to the $(t-1)^{\text{th}}$ hidden state and finally back to the input data (again, see Eq. 5.1).

Remember that our objective is to update the weights of the RNN model. Let's pick, for instance, W_{ax} and work out the **chain rule** to go from $\frac{\partial L}{\partial \hat{y}^{<t>}}$ to $\frac{\partial L}{\partial W_{ax}}$: this entails moving backwards along the network, as illustrated in Fig. 5.5 for position <t> in the sequence (the last time step).

In a more formal mathematical notation, the chain rule to get the partial derivative of the loss function with respect to the target weights W_{ax} is the following:

$$\frac{\partial L}{\partial W_{ax}} = \frac{\partial L}{\partial \hat{y}^{<t>}} \cdot \frac{\partial \hat{y}^{<t>}}{\partial a^{<t>}} \cdot \frac{\partial a^{<t>}}{\partial a^{<t-1>}} \cdot \frac{\partial a^{<t-1>}}{\partial W_{ax}} \tag{5.2}$$

This was the gradient of the input weights (W_{ax}) for a single time step: in a typical RNN, though, we have multiple elements in the sequence, and Fig. 5.5 (top-right corner) shows that the overall loss of the network is the sum of the losses of all the single elements in the sequence. How does this reflect into the calculations

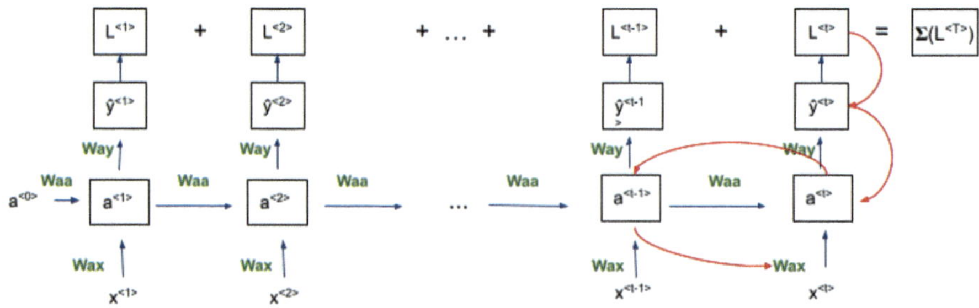

Fig. 5.5 Backpropagation through time: visual representation of the chain rule to get the partial derivative of the loss function with respect to the input weights (weights to update) for a single time step (position/step <t> in the sequence—the last one)

of partial derivatives involved in BPTT? Since the losses are summed together, we should also sum the gradients of the individual time steps: however, since the time steps are interconnected through the hidden states (the activations), we must somehow take this into account.

If we consider the last two time steps (remember, in backpropagation we move backwards, from the output of the network to the input), the BPTT calculations to obtain the updated input weights W_{ax} become:

$$\frac{\partial L}{\partial W_{ax}} = \frac{\partial L}{\partial \hat{y}^{<t>}} \cdot \frac{\partial \hat{y}^{<t>}}{\partial a^{<t>}} \cdot \frac{\partial a^{<t>}}{\partial a^{<t-1>}} \cdot \frac{\partial a^{<t-1>}}{\partial W_{ax}} + \frac{\partial L}{\partial \hat{y}^{<t-1>}} \cdot \frac{\partial \hat{y}^{<t-1>}}{\partial a^{<t>}} \cdot \frac{\partial a^{<t>}}{\partial a^{<t-1>}} \cdot \frac{\partial a^{<t-1>}}{\partial a^{<t-2>}} \cdot \frac{\partial a^{<t-2>}}{\partial W_{ax}} \tag{5.3}$$

As you can see in Eq. 5.3, for the second-to-last element in the sequence we have one additional term in the chain, specifically the next hidden state (going backwards) <t-2>. This is actually the hidden state corresponding to the way in which the output at <t-1> is predicted, which involves the activation values from the previous time step (Eq. 5.1): however, when moving backwards, since the elements in the sequence are interconnected, we also have to consider all the activation values from the "future", in this case those at <t>. Moving backwards along the sequence, the combined hidden states accumulate progressively, until they will include the whole sequence.

This can be expressed mathematically in a concise and elegant way:

$$\frac{\partial L}{\partial W_{ax}} = \sum_{i=0}^{T} \frac{\partial L}{\partial \hat{y}^{<t-i>}} \cdot \frac{\partial \hat{y}^{<t-i>}}{\partial a^{<t>}} \left(\prod_{j=T-i+1}^{T} \cdot \frac{\partial a^{<t-j+1>}}{\partial a^{<t-j>}} \right) \cdot \frac{\partial a^{<t-i-1>}}{\partial W_{ax}} \tag{5.4}$$

All right, we got the gradients for the input weights W_{ax}. In the very same way we can calculate the gradients for the recurrent weights W_{aa}, just by taking the partial derivative of the loss function with respect to W_{aa} instead:

$$\frac{\partial L}{\partial W_{aa}} = \sum_{i=0}^{T} \frac{\partial L}{\partial \hat{y}^{<t-i>}} \cdot \frac{\partial \hat{y}^{<t-i>}}{\partial a^{<t>}} \left(\prod_{j=T-i+1}^{T} \cdot \frac{\partial a^{<t-j+1>}}{\partial a^{<t-j>}} \right) \cdot \frac{\partial a^{<t-i-1>}}{\partial W_{aa}} \tag{5.5}$$

The derivation of the loss function with respect to the output weights W_{ay} is a bit simpler: the reason for this is that the output weights—similar to the other sets of weights—are shared across the time sequence and we can therefore differentiate for them at each time step and then sum everything together, but they are not recurrent and therefore there is no need for the multiplication term $\prod$. Here's how the gradients for the output weights are calculated:

$$\frac{\partial L}{\partial W_{ay}} = \sum_{i=0}^{T} \frac{\partial L}{\partial \hat{y}^{<t-i>}} \cdot \frac{\partial \hat{y}^{<t-i>}}{\partial W_{ay}} \tag{5.6}$$

We are left only with the gradients for the **bias terms**:

$$\frac{\partial L}{\partial W_{ba}} = \sum_{i=0}^{T} \frac{\partial L}{\partial \hat{y}^{<t-i>}} \cdot \frac{\partial \hat{y}^{<t-i>}}{\partial a^{t-1}} \cdot \frac{\partial \hat{a}^{<t-i>}}{\partial b_a} \tag{5.7}$$

$$\frac{\partial L}{\partial W_{by}} = \sum_{i=0}^{T} \frac{\partial L}{\partial \hat{y}^{<t-i>}} \cdot \frac{\partial \hat{y}^{<t-i>}}{\partial b_y} \tag{5.8}$$

Once we have all the partial derivatives for the weights and bias terms in Eq. 5.1, we can proceed to the **updating step** before the next forward propagation pass. To update the weights and bias terms, we do what we have already seen for dense neural networks in Eq. 4.6: we multiply the partial derivative of the loss function with respect to the coefficients by the **learning rate** α, and subtract this from the current values for the coefficients (remember: if the slope is positive, we need to reduce the coefficient value, if it's negative, we do the opposite—see Fig. 4.7). As an example, for the input weights the updating works as in Eq. 5.9 (similar expressions are used for all other coefficients—not shown here):

$$W_{ax} := W_{ax} - \alpha \cdot \frac{\partial L}{\partial W_{ax}} \tag{5.9}$$

To sum up, **Backpropagation through time** (BPTT) is an adaptation of the basic algorithm that updates the weights across the different "time steps" (positions) of data elements in a sequence model. In this way, RNN models are able to capture temporal (order) dependencies and patterns in sequence data (remember: order matters!). This is achieved through the implicit "memory" of recurrent connections (**recurrent weights**) that allow RNN models to learn from previous (past) inputs in the sequence: with BPTT the model parameters are adjusted (updated) factoring in the order in the sequence.

The basic RNN model and BPTT algorithm described up to this point are an excellent starting point, but they tend to suffer quite dramatically from the **vanishing gradients** problem (see box "Vanishing gradients" in Sect. 4.3). Fortunately, we have some solutions, and we'll mention them in one of the next sections, after a brief digression on a few specialised RNN architectures.

5.4 Specialised RNN Architectures

In Fig. 5.3 we described a simple unidirectional RNN model for the analysis of sequence data (a more accurate representation of this model can be found in Fig. 5.4, but for the purpose of this illustration we'll stick to the flattened visualization of RNNs). This specific type of RNN arrangement is called **many-to-many architecture**: many inputs ($x^{<t>}$) are mapped to **as many outputs** ($y^{<t>}$); examples include the labelling of each input word in a text into categories (e.g. grammar tokens as adjectives, nouns, verbs etc.), or the transcription of spoken words into the corresponding written counterparts.

However, not all problems fall into this category, in which the exact same number of inputs corresponds to the exact same number of outputs. In some (many) sequence-data problems, we may want to classify whole text data (e.g. newspaper articles) according to topics (e.g. sport, politics, economy etc.). In generative models, a single input (e.g. type of music) may be used to produce a whole sequence of output data (e.g. musical notes). Machine translation models face yet another different type of problem, where an input sequence of length n should be mapped to a translated output sequence of length m. These types of problems have spurred the development of specialised RNN architectures: below we list a few such architectures, which can be visualised (flat representation) in Fig. 5.6.

- **many-to-one** architecture: e.g. read a reviewer report and understand whether it recommends to accept, reject or revise a submitted scientific article
- **one-to-many** architecture: e.g. choose the type of short story to write (mystery, romance, drama etc.) and generate the corresponding output text
- **generalised many-to-many** architecture: mapping input and output sequences with different lengths (e.g. machine translation, or making a summary of an input sequence)

We mentioned only a few RNN architectures, but there is virtually no limit to how RNN units can be creatively combined to tackle all possible problems involving sequence data and a longitudinal (time) component (e.g. bidirectional RNNs, hierarchical RNNs, attention-based RNNs, clockwork RNNs etc.). For practical implementation and examples see the notebook in Sect. A.12.

On top of this, we have so far discussed what happens inside **one single layer** of a RNN: however, we can of course stack together several RNN layers, to build a **deep RNN** (like like things weren't complex enough! See Fig. 5.7). A deep RNN would generalise the forward propagation calculations that we saw in Eq. 5.1. Let's take as an example the red-circled cell in Fig. 5.7: this is the second layer ([2]) at the third time step (<3>). The recurrent activation values will be, as usual, those of the same layer ([2]) but from the previous time step (<2>); on the other hand, the input in this case will be the activation values from the same time step (<3>), but from the previous layer (1).

In a notation a bit more compact than that in Eq. 5.1, we have:

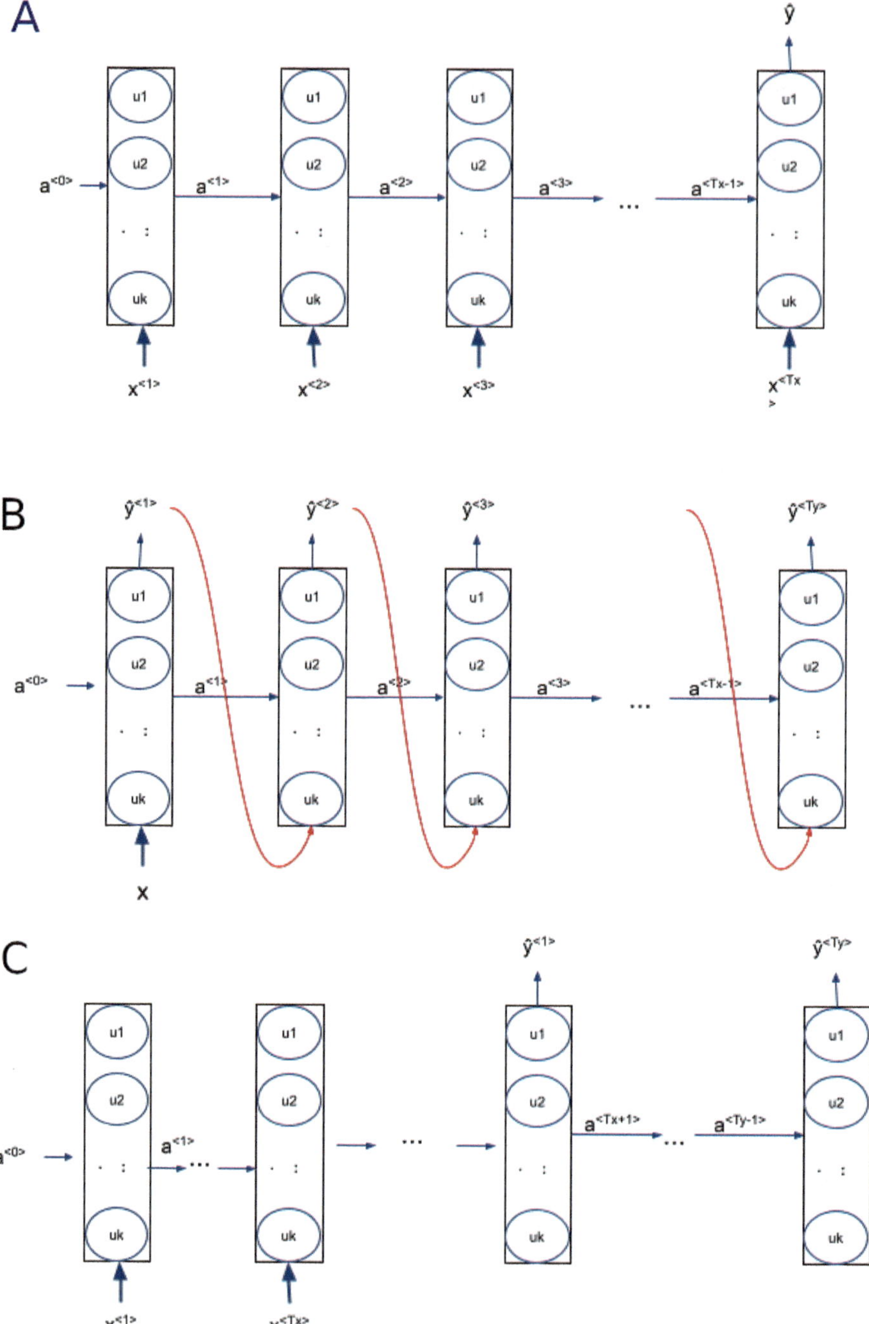

Fig. 5.6 A few specialised RNN architectures: (**a**) many-to-one: sequence input to predict one target; (**b**) one-to-many: single input to generate a sequence output the "prediction" at each time step <t> is the input for the next time step <t+1> (together with the activation values); (**c**) generalised many-to-many: sequence inputs and outputs, of different lengths; here the first part of the network only encodes the input data (no output), while the second part of the network 'decodes' the encoded data

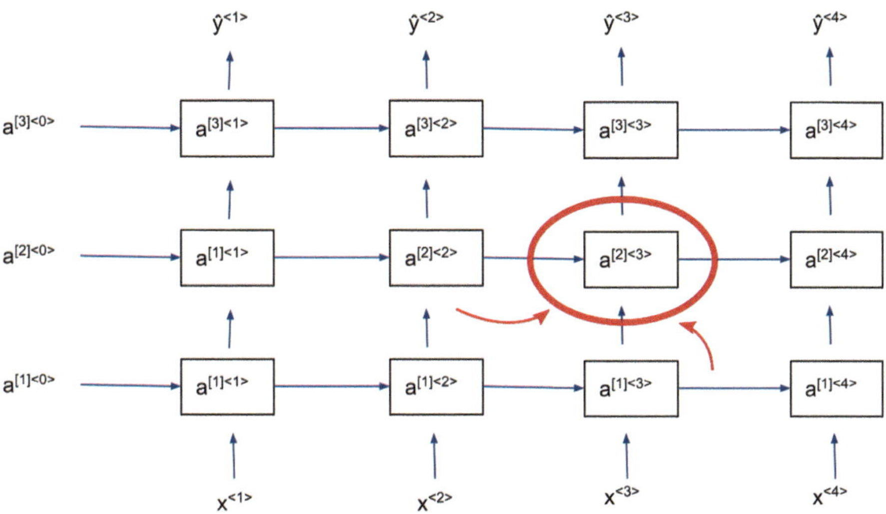

Fig. 5.7 Deep recurrent neural network (RNN). In evidence (circled in red): the cell in the second layer ([2]) at the third time step (<3>)

$$a^{[2]<3>} = g\left(\mathbf{W}_{\mathbf{a}}^{[2]} \cdot \left[a^{[2]<2>}, a^{[1]<3>}\right] + b_a^{[2]}\right) \tag{5.10}$$

where $\mathbf{W_a}$ is a matrix of weights that stacks together—side by side—the two matrices $\mathbf{W_{aa}}$ for input $a^{[2]<2>}$ (same layer, previous time step), and $\mathbf{W_{aa}}*$ for input $a^{[1]<3>}$ (previous layer, current time step). This matrix will have the following dimensions: $W_{a(u,u+u*)}$, where u and $u*$ are the number of units from the two layers. Obviously, the two input vectors will also be stacked together, one on top of the other, with dimensions: $\left[a^{[2]<2>}, a^{[1]<3>}\right]_{(u+u*,1)}$.

Embeddings: A New Representation for Sequence Data

In Fig. 5.1 we introduced **one-hot encoding** of large vocabularies to represent sequence data for RNN models. We can easily see that this type of **data representation** has one obvious drawback: it will soon lead to huge and very sparse data matrices, as the number of terms in the vocabulary increases. This can be suboptimal and inefficient.

A solution to this problem is provided by **embeddings**. Embeddings are an alternative data representation for sequence data that replaces large sparse matrices of 0s and 1s with more compact **lower-dimensionality dense matrices** of numerical values. These numerical values are—ça va sans dire—learnt autonomously by the neural network, and are not interpretable by the user/developer (this is an actual black box).

(continued)

term (input data)	mammal	female	male	It flies
mare	0.89	0.94	0.01	0.01
bull	0.90	0.001	0.99	-0.302
hen	-0.0025	0.984	0.0231	0.257
eagle	0.0301	0.3056	0.519	0.879

Fig. 5.8 Simplified intuition of embeddings for the encoding of sequence data

To provide some intuition, we can have a look at the simplified example in Fig. 5.8: here we have, in the first column of the table, four terms (input data) that in the conventional OHE representation would be encoded as very long vectors of 0s and (a single) 1, with dimensionality p equal to the size of the vocabulary, say the 42,000 words known by an average 20-year-old native speaker of American English [89]. With embeddings, the same four input data will be compressed into a lower dimensionality representation, in our example to just four variables: being a mammal, being a female, being a male, the ability to fly. Numbers will be assigned to each of these new variables, indicating how much of each corresponds to the input data and in which direction (the sign). For example, a mare will have high positive values for being a mammal and a female, but values close to zero for being a male and for flying (the existence of the expression "mare's nest" may prevent this term from having negative values for flying). Conversely, the hen will have a low value for being a mammal, and maybe a moderate value for flying. The term "eagle" will also have low values for being a mammal, but high values for flying, while it is not clear whether this refers to a male or a female (hence the moderate values).

Keep in mind that these are made-up numbers used to provide the reader with some intuition of what embeddings are. In reality, the neural network will create $p_{embeddings}$ new variables (var_1, $var_2 \ldots var_p$, with $p_{embeddings} << p$) that are not interpretable, but work spectacularly well in condensing the information contained in the data into a form that is highly effective for the purpose of the model (prediction, classification, clustering, generation of content etc.).

Although embeddings are most intuitively applied to sequences of text data, they can also be used for other types of data like images, sounds, etc. (e.g. [90]).

5.5 Memory Pills: LSTM and GRU Units

In the box "Vanishing gradients" (Part 1: "Neural networks and deep learning decoded"), we introduced the issue of **vanishing gradients**, which is a general problem that affects deep neural networks models: as the number of layers increases, it gets progressively more difficult for forward and—especially—backpropagation to transfer the learning across the network.

In a deep neural network the backpropagation algorithm has difficulties exerting an effect (i.e. updating the weights) on the early layers of the network. A basic RNN is indeed a sort of deep neural network, as the elements of the sequence are processed sequentially, and the activation weights are supposed to store information and transfer learning over long distances, from the beginning to the end of the sequence. The basic RNN model is not particularly good at capturing distant relationships in the sequence, and this can be thought of as a problem of "memory loss". To give an example, in the sentence below, the network needs to remember whether the subject is singular or plural (third word) to decide which verb to use (`is` or `are`) at the end of the sentence (second-last word). The basic RNN model will tend to forget the noun "number" (singular/plural), thereby failing to correctly model the long sentence.

Listing 5.1 Example of a long sentence

```
1 The red cow, living in the woods, grazing on ... and
      mating during ... at temperate latitudes ... is a
      ruminant
2 The red cows, living in the woods, grazing on ... and
      mating during ... at temperate latitudes ... are
      ruminants
```

As discussed in the box "Vanishing gradients", to provide some intuition as to why vanishing gradients happen, with a little simplification we can think of a neural network as a series of matrix products: $\mathbf{W}^L$, where $\mathbf{W}$ is a matrix of weights and L is the number of layers. Since weights in a neural network tend generally to be small numbers between -1 and $+1$ (normalised input data, shrinkage of coefficients due to the many parameters usually involved in a neural network model), with progressive multiplications (exponentiation) they tend to get smaller and smaller as the number of layers increases. This leads to the vanishing gradient problem: from Eq. 4.6 we know that the new weight value is given by the old weight value plus the **product of the learning rate and the gradient**. If the gradient is very small, we may have for instance something like $0.75194 = 0.75195 - 0.1 \cdot 0.0001$: not much of a difference between the old value and the updated value of the weight. In such cases, the neural network can get stuck and not be able to make any progress in learning the problem it is supposed to model and solve.

To solve this problem, "memory pills" for RNN models have been invented, in the form of specialised units and layers: **GRU** (gated recurrent units) and **LSTM** (long-short term memory) units. To provide some intuition, these specialised units do something that we often do as humans: when we read something or listen to someone speaking, we do not take note of and remember all the words that we read or hear, but we rather focus on the ones that are important to us in understanding the key message. For example, when reading the instructions for a new appliance we probably subconsciously forget (or directly skip!) technical details and alternative options, and focus only on the key points to get it to work. GRU and LSTM units do something similar: they retain the important terms of the sequence to be used later on to make decisions (remember, the neural network will autonomously learn what is important to it, that's the job of the network!).

From an historical perspective, LSTM units and layers were invented in 1997 [91], long before GRU layers, which came about in 2014 and are a clever refinement of previous attempts at modelling long-range dependencies in sequence data [92]. However, GRU layers are simpler and, for the purpose of this illustration, we will start from the description of GRU to then move on to LSTM.

GRU Units: To Update or Not to Update, That Is the Question

In Fig. 5.9 we see the sketches of a basic RNN unit (left) and of a **GRU unit** (right). In the basic RNN unit, the input data $x^{<t>}$ and the activation values from the previous time step $a^{<t-1>}$ are weighted by their respective coefficients and combined to produce the new activation values $a^{<t>}$ (later used to predict $\hat{y}^{<t>}$ and passed to the next time step in the sequence). This is illustrated mathematically in Eq. 5.1.

GRU stands for **gated recurrent unit** and improves on the basic RNN unit by using the **memory cell** $c^{<t>}$, which is another set of values (like the activations values) that are used to remember elements of the sequence. Actually, in GRU layers memory cell values are exactly equal to the activation values: $a^t = c^t$ (they are the same thing, as far as GRUs are concerned).

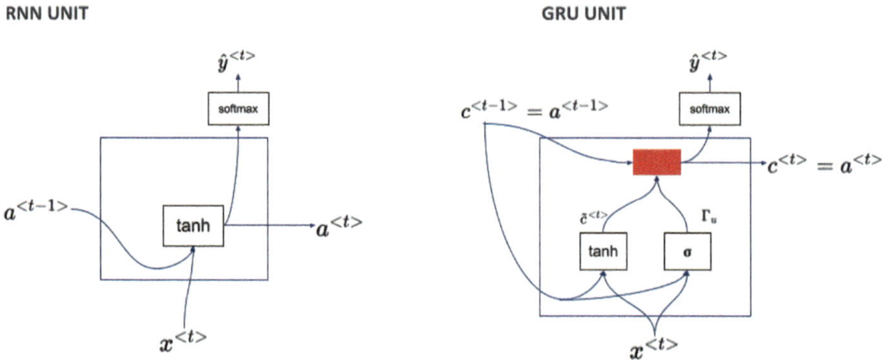

Fig. 5.9 Basic RNN unit (left) and GRU unit (right). In the red box the following algebraic operations take place: $c^{<t>} = a^{<t>} = \Gamma_u \cdot \tilde{c}^{<t>} + (1 - \Gamma_u) \cdot c^{<t-1>}$

New memory cell values at the current time step are "potential new values" (the updating of the values is only potential), are indicated as $\tilde{c}^{<t>}$, and are calculated similarly to activation values in basic RNN units, with their specific set of weights W_{cc} (analogous to W_{aa} in Eq. 5.1):

$$\tilde{c}^{<t>} = \tanh\left(W_{cc} \cdot c^{<t-1>} + W_{cx} \cdot x^{<t>} + b_c\right) \tag{5.11}$$

Now we get to the key question: will the memory cell (activation) values be updated or not? This is the key aspect of GRU layers, and is controlled by the **update gate** Γ_u, which returns a scalar based on previous memory-cell values (activations) and current input data, all weighed by yet another two sets of trainable parameters (W_{uu}, W_{ux}):

$$\Gamma_u = \sigma\left(\mathbf{W_{uu}} \cdot c^{<t-1>} + \mathbf{W_{ux}} \cdot x^{<t>} + b_c\right) \tag{5.12}$$

When we combine Eqs. 5.11 and 5.12 we get the following **update rule**:

$$c^{<t>} = \Gamma_u \cdot \tilde{c}^{<t>} + (1 - \Gamma_u) \cdot c^{<t-1>} \tag{5.13}$$

This (the calculations in Eq. 5.13) is what happens in the red box of Fig. 5.9 (right), and it is the crux of the GRU unit: if Γ_u is close to one, the memory-cell values will be updated ($\tilde{c}^{<t>}$ will replace $c^{<t>}$); otherwise, if Γ_u is close to zero, the previous memory-cell values will be left untouched ($c^{<t-1>}$ will be passed on to become $c^{<t>}$, without further processing).

Remember ("repetita juvant"): it is the job of the GRU layer (i.e. layer made of GRU units) to learn when to update the memory-cell values and by how much, or for how long to keep the old values in its memory. It does this by learning (i.e. setting the values) of the many sets of weights we encountered in Eqs. 5.11 and 5.12. The update gate will decide how much information from the previous time step to keep ($c^{<t-1>} = a^{<t-1>}$) and how much information from the current step to gather. This is very powerful, since the GRU unit can decide—through the update gate—to conserve all past information and ignore the new information, thereby reducing the risk of vanishing gradients (if $\Gamma_u \approx 0 \rightarrow c^{<t>} = c^{<t-1>}$, this reduces the number of matrix multiplications through the network: see the box "Vanishing gradients").

To provide some intuition, let's again take the "cow" example sentence from Sect. 5.5: the job of the network is to remember whether the subject is singluar ("cow") or plural ("cows"), and then choose the right verb form accordingly ("is" or "are"), several words down the line. We can think that the value for the update gate Γ (from Eq. 5.13) is set to one when memory cell values (activations) need to be updated because something important to be remembered is found in the sequence (for example the fact that "cow" is singular). Γ is then set to zero to avoid updating—and thus remembering! (the information stored in memory is passed forward unscathed)—further down along the sequence, until the next important bit is encountered (e.g. the matching-case verb "is") and the memory cell values are again "free" to be updated to remember something new (Fig. 5.10).

Γ = 0 0 1 0 ... ... 0 ... ... 0 ... 1 0 0

The red cow, living in the woods ... and mating during ... at temperate ... is a ruminant

Fig. 5.10 Intuition on how memory cells work. The update gate Γ is set to one when the memory cell needs to be updated because something to be remembered is found in the sequence. It is then set to zero to avoid updating—and thus remembering!—along the sequence until the next important bit is encountered (and the memory cell values can now remember something new—if needed)

Actual GRU implementations are a little more sophisticated than what we illustrated, and usually contain a reset gate alongside the update gate e.g. [93]. Furthermore, the memory cell $c^{<t>}$ and update gate Γ can be vectors rather than scalars to remember multiple things at the same time (e.g. singular/plural, past/present, he/she etc.).

GRU layers are good at learning when to update $c^{<t>}$ and then keep it constant until used, and by balancing the reset and update gates GRUs are able to effectively capture short-term and long-term dependencies in the sequence.

LSTM Units

As mentioned, LSTM units were developed earlier than GRU units, but present a more complex architecture and therefore we decided to illustrate GRU units first. We can think of GRU as a refinement of LSTM units, that achieve similar learning performances by doing fewer calculations (thereby being more efficient). This said, GRU and LSTM layers are both state-of-the-art components of modern RNN models for sequence data.

Let's now have a look at the LSTM unit (Fig. 5.11): the first thing we notice is that here the activation values $a^{<t>}$ are distinct from the memory cell values $c^{<t>}$ (unlike GRU units, where they are collapsed). Besides, we see that in terms of gates, LSTM units feature the following: (i) **update gate** Γ_u; (ii) **forget gate** Γ_f; and (iii) **output gate** Γ_o.

These gates take care of learning what to remember and forget when reading the sequence of data, in a similar—albeit slightly more complex—way as what we described for GRU layers. As you may have guessed, all these gates do not come for free, but have ther associated sets of **trainable parameters** $\mathbf{W_{uu}}$, $\mathbf{W_{ux}}$, $\mathbf{W_{ff}}$, $\mathbf{W_{fx}}$, $\mathbf{W_{oo}}$ and $\mathbf{W_{ox}}$. The accompanying notebooks in Sects. A.11 and A.12 illustrate the number of trainable parameters to be learnt in RNN models with basic, GRU and LSTM units.

Figure 5.12 provides a summarised comparison of the **tensor (matrix) calculations** involved in GRU and LSTM units. We can see that by collapsing together activations ($a^{<t>}$) and memory cell values ($c^{<t>}$), and by getting rid of one gate (two vs three gates), GRU units considerably reduce the number of calculations to be performed, thereby improving their relative efficiency with respect to LSTM units.

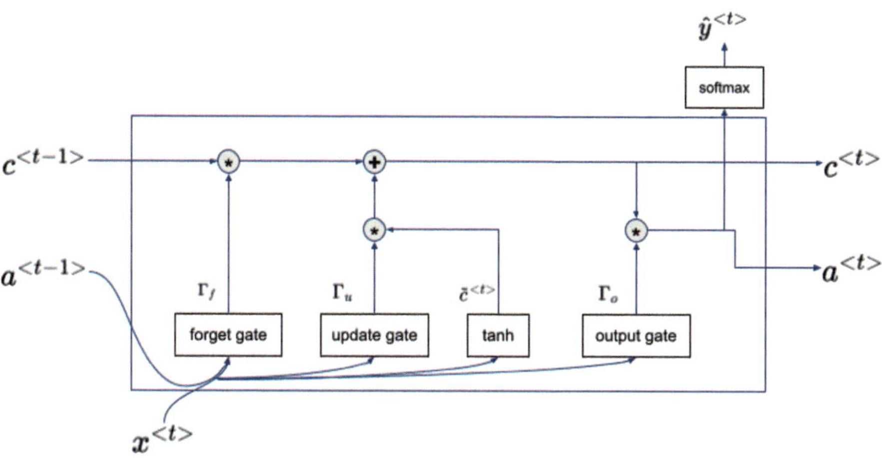

Fig. 5.11 Schematic representation of a LSTM (long-short-term memory) unit: $x^{<t>}$ are input data, $c^{<t>}$ are memory cell values, $a^{<t>}$ are activation values, $\hat{y}^{<t>}$ are predictions, Γ_u, Γ_f and Γ_o are the update, forget and output gates. The + and * signs in the circles indicate summations and multiplications

GRU unit vs **LSTM unit**

GRU unit	LSTM unit
$\tilde{c}^{<t>} = \tanh\left(W_c[\Gamma_r \cdot c^{<t-1>}, x^{<t>}] + b_c\right)$	$\tilde{c}^{<t>} = \tanh\left(W_c[a^{<t-1>}, x^{<t>}] + b_c\right)$
$\Gamma_u = \sigma\left(W_{uu} \cdot c^{<t-1>} + W_{ux} \cdot x^{<t>} + b_u\right)$	$\Gamma_u = \sigma\left(W_{uu} \cdot a^{<t-1>} + W_{ux} \cdot x^{<t>} + b_u\right)$
$\Gamma_r = \sigma\left(W_{rr} \cdot c^{<t-1>} + W_{rx} \cdot x^{<t>} + b_r\right)$	$\Gamma_f = \sigma\left(W_{ff} \cdot a^{<t-1>} + W_{fx} \cdot x^{<t>} + b_f\right)$ $\Gamma_o = \sigma\left(W_{oo} \cdot a^{<t-1>} + W_{ox} \cdot x^{<t>} + b_o\right)$
$c^{<t>} = \Gamma_u \cdot \tilde{c}^{<t>} + (1 - \Gamma_u) \cdot c^{<t-1>}$	$c^{<t>} = \Gamma_u \cdot \tilde{c}^{<t>} + \Gamma_f \cdot c^{<t-1>}$
$a^{<t>} = c^{<t>}$	$a^{<t>} = \Gamma_o \cdot \tanh(c^{<t>})$

Fig. 5.12 Summary comparison of the tensor (matrix) calculations that need to be performed in GRU vs LSTM units (Γ_r indicates the reset gate)

What Goes on Inside the Units (Nodes) of Neural Networks

In Sect. 4.2 we saw that inside a DNN (deep neural network) unit the input variables are multiplied by their respective weights and summed together; finally, a bias term is added: $\sum_{j=1}^{p}(w_j \cdot x_j) + b$.

This is a linear combination of the input variables, i.e. **linear regression**.

(continued)

But do DNN units always compute linear regression? By now we are sure that you'll guess the answer: no, they do not!

In **CNN** (convolutional neural network), we saw that units calculate **convolutions**, which are **Hadamard products** between matrices followed by summations. Yet more, in simple **RNN** the units calculate a linear combination of the input features **plus the activations from the previous time step** ("from the past"). This is further complicated by **memory cells**, **update**, **reset**, **forget** and **output gates** in GRU and LSTM units, as we described in Sect. 5.2 and in "GRU units: to update or not to update, that is the question".

Now you got the gist of it: in deep learning what you need to do is come up with ("invent") some type of sets of calculations, then repeat them thousands of times, let the neural network learn the parameters and "bam" ...the job is done!

Part III

Making It Work

6 Making It Work

After describing what deep learning and neural networks are, as well as the main architectures available for the most common DL problems (classification, regression, image recognition, sequence data) it is now time to discuss some practical aspects that have to be faced when moving from the neat and tidy world of theory to the practical applications of DNNs. These practical aspects include: the preprocessing of input data ("getting the data right"), some techniques for increasing the efficiency of neural networks, additional metrics for evaluating the performance of DNN models, more advanced DNN models and novel types of problems to tackle, general ways of setting up a functioning deep learning project, and methods for visualization and extraction of feature importance from neural networks ("explainable AI"). Let's get started!

6.1 Preparing the Input Data: Preprocessing and Augmentation

First things first: getting the data right is obviously the necessary initial operation when you want to work with machine/deep learning. After cleaning and filtering (operations that are very specific to each dataset, and are outside the scope of this book), data are normalised before entering the neural network. Shall we also transform the data, e.g. the target variable, under some circumstances? And what about the fancy-sounding idea of artificially increasing the size of our data? Let's discuss!

Data Normalization

Data normalization is the process of transforming your input data so that it becomes better suited for neural networks and machine-learning algorithms in general.

F. Biscarini, N. Nazzicari, *Deep Learning for Life Sciences*, Decoding Evolution 1,
https://doi.org/10.1007/978-3-031-96852-5_6

There are some statistical and machine-learning methods that absolutely require the input data to be normalised (e.g. principal component analysis, Lasso-penalised regression, support vector machines depending on the chosen kernel: basically, all methods where the different variables are directly compared); for some other methods, on the other hand, it is not strictly necessary to normalise the input data (e.g. tree-based methods like Random Forest, multiple linear or logistic regression). In general, however, if your objective is to make predictions and not to estimate some parameters of specific X variables, even when not strictly necessary, it is not harmful—and usually a good practice—to normalise the input features.

Neural networks belong to those methods for which the **normalization of the input data is necessary**: remember, neural networks work by assigning **weights** to the input features in the first and subsequent layers, and if features are on different scales some of them will matter more (i.e. have higher weight) not because they are more relevant but simply because their scale (range, mean, standard deviation) is larger. Hence, the requirement for the normalization preprocessing step.

It intuitively makes sense to normalize tabular data; that's the approach most of us are already familiar with. However, even image data (and other types of data: text, videos, sounds etc.) benefit from (and must be subjected to) normalization (for gray-scale raster images, for example, pixel intensities in the range 0–255 are normalised to be in 0–1). Data normalization typically works by making data more homogeneous, and thus more palatable (easier to process) for the network. Imagine you want to train a network to predict if some patients suffer from a heart disease. The input data may look like what's reported in Table 6.1. Just looking at the table without getting into the details, it is clear that the data could use a bit of normalization: some columns/features (e.g. *trestbps*) are numbers in the hundreds, others (e.g. *age*) in the tens, and yet others (e.g. *oldpeak*) in the ones (units) or less. This variety of orders of magnitude makes the job of the neural network harder. Remember that each row of the table is taken as input by each unit of the first layer of the network, which will need to compensate for the differences in data size with appropriate coefficients so that the larger variables do not completely obfuscate the smaller ones. And that must happen *on top* of the regular learning process. Luckily, the issue can easily be fixed before presenting the data to the network by normalizing each column (each feature) separately.

There are several approaches to data normalization. The two most popular are:

Min-Max Normalization This approach imposes that the normalized data live within a defined range, typically between zero and one. Given a set of input data $x_1, x_2, \ldots, x_n$, the normalized series is defined as:

$$x_i' = \frac{x_i - x_{MIN}}{x_{MAX} - x_{MIN}}$$

where x_i' is the normalized version of data point x_i and x_{MAX} and x_{MIN} are the series maximum and minimum values, respectively.

Table 6.1 An example of tabular data that would benefit from normalization: the Cleveland heart disease dataset. Here for brevity we report only the top five rows (i.e. patients), which have a heart condition (*target*=1) and the bottom five rows, which correspond to healthy patients (*target*=0). Columns represent, in order: *age* (years); *sex* (1 = male; 0 = female); *cp*: chest pain type (1 = typical angina, 2 = atypical angina, 3 = non-anginal pain, 4 = asymptomatic); *trestbps*: resting blood pressure (in mm Hg on admission to the hospital); *chol*: serum cholesterol in mg/dl; *fbs*: (fasting blood sugar > 120 mg/dl) (1 = true, 0 = false); *restecg*: resting electrocardiographic results (0 = normal, 1 = having ST-T wave abnormality, 2 = probable or definite left ventricular hypertrophy), *thalach*: maximum heart rate achieved; *exang*: exercise induced angina (1 = yes, 0 = no); *oldpeak*: ST depression induced by exercise relative to rest; *slope*: the slope of the peak exercise ST segment (1 = upsloping, 2 = flat, 3 = downsloping); *ca*: number of major vessels (0–3) colored by fluoroscopy; *thal* (3 = normal, 6 = fixed defect, 7 = reversible defect). Data from: https://www.kaggle.com/datasets/johnsmith88/heart-disease-dataset

Age	Sex	cp	trestbps	chol	fbs	restecg	thalach	exang	oldpeak	Slope	ca	thal	Target
63	1	3	145	233	1	0	150	0	2.3	0	0	1	1
37	1	2	130	250	0	1	187	0	3.5	0	0	2	1
41	0	1	130	204	0	0	172	0	1.4	2	0	2	1
56	1	1	120	236	0	1	178	0	0.8	2	0	2	1
57	0	0	120	354	0	1	163	1	0.6	2	0	2	1
57	0	0	140	241	0	1	123	1	0.2	1	0	3	0
45	1	3	110	264	0	1	132	0	1.2	1	0	3	0
68	1	0	144	193	1	1	141	0	3.4	1	2	3	0
57	1	0	130	131	0	1	115	1	1.2	1	1	3	0
57	0	1	130	236	0	0	174	0	0	1	1	2	0

Z-Score Normalization This method standardizes the data by forcing it to have zero mean and unitary standard deviation. The transformed data is defined as:

$$x'_i = \frac{x_i - \mu}{\sigma}$$

where x'_i is the normalized version of data point x_i, μ is the mean of the data, and σ is the standard deviation.

The notebook in Sect. A.6 shows an example of input data normalization and of learning trajectories with and without normalization.

Data Transformation

Besides being normalised, data can be transformed so as to change their shape (normalisation is invariant to the distribution of the data). The reason is usually that the original data may come in a form that is far from the comfort zone of Gaussian (bell-shaped) distribution and linear additive relationship between y and x (or multiple x's).

Therefore, mathematical transformations can be applied to: (i) convert a skewed distribution to a Gaussian distribution; (ii) linearize the relationship between y and X; (iii) reduce heteroscedasticity (heterogeneity of variance); (iv) produce a clearer visualization of the data.

Transformations are typically applied to the **target variable** y, while it is less common to transform the predictors X, since there are often many of these and they may require different types of mathematical operations. Popular transformations of the target variable include: (i) **logarithms** (usually the natural or base-10 logarithm); (ii) **square root**; (iii) **Box-Cox power transformation.**[1] These transformations are mathematically defined for positive numerical values, and some preprocessing is required with negative and zero-valued variables (e.g. rescaling).

It is not uncommon to find data transformation in the bag of tricks of classical statistics. However, it is not really useful or applicable (nor it is applied) in the context of machine-learning, neural networks and deep learning, mainly because the types of data analysed by DNN models tend not to be tabular data, but rather image, sequence, video or text data. Therefore, we will not delve further into the rather large topic of data transformations: if you wish to know more on this, you can refer to the review by Fink [95] and to Chapter 10 of the book by Reimann et al. [96]), both of which are excellent starting points.

6.2 Data Augmentation

Data augmentation may sound like something totally new and specific to deep learning. This is not entirely true, as approaches to artificially changing the size of the original dataset have been devised to deal with the problem of imbalanced data in binary (or multinomial) classification problems. The simpler way is to reduce the size of the majority class by randomly removing some examples, so that both (or all) classes are of similar size (**downsampling**). Another possibility is to increase the size of the minority class, by generating new synthetic examples in its feature space (**upsampling** or **oversampling**): examples of such methods are **SMOTE** (Synthetic Minority Oversampling TEchnique [97]) and **ROSE** (Randomly Over Sampling Examples [98]).

More specific to deep learning, **data augmentation** is a general technique used to **artificially increase the size of the training dataset**. It starts from the following observations:

[1] In the most common form, when $\lambda \neq 0$ [94] the Box-Cox transformation looks like:

$$y_i^{(\lambda)} = \frac{y_i^{\lambda} - 1}{\lambda \mathrm{GM}(y)^{\lambda - 1}}$$

(GM is the geometric mean).

- almost always, more data means more accuracy. Anything that allows for larger datasets should be considered
- all datasets are noisy. Repeating the measures on exactly the same samples will result in similar—but not identical—numbers. We want the network to follow the patterns and ignore the noise
- any piece of data is collected under a specific condition. Maybe the sensor is very close (or very far) from the source. Maybe the camera is tilted. Maybe the background is patchy. We want the network to generalize the features of the subject and forget about any systematic features in the dataset

Data augmentation uses the training data as a template to generate more data examples. In general terms, the original data are transformed in random ways, either by adding noise or by leveraging other types of transformations. Pay attention: only the training data is augmented, and the golden rule holds. You can do anything you want on the training dataset as long as you leave the validation dataset unscathed.

The specific way by which the new, augmented dataset is generated depends on the type of data we are dealing with, the most common case being that of image data.

In general, data augmentation works because it exposes the model to a wider variety of examples during training, which helps the network to separate noise from the signal and to learn more robust features that can generalize better to new, unseen data. This makes the model more robust and less likely to overfit to the specific characteristics of the training data.

Data Augmentation for Images

By far, the most common application of data augmentation is to image data. Images can be easily manipulated in many sensible ways. The core observation is that many image transformations don't alter the nature of the subject, while they can quite radically change the numeric values (the pixels) involved (Fig. 6.1).

Depending on the scientific problem at hand, specific types of image transformations will make sense (or not). The list of possible ways of augmenting an image dataset is potentially unbounded, but in practical terms your deep learning library of choice will have some built-in capabilities. The most common transformations for the augmentation of image data are:

- flip: horizontally, vertically or both
- shift: the image is moved (e.g. 50 pixels to the right) so that part of it goes beyond the border and an empty rectangle appears on the opposite side
- rotation: the image rotates, clockwise or counterclockwise. Four triangular areas go outside of the border and four new empty triangular areas appear
- shear: the image is deformed as if the top border and the bottom border were pushed in opposite directions
- zoom in, zoom out

Fig. 6.1 Why data augmentation works. All the pixels of the left image are different from all the pixels of the right image. From the numerical point of view they are completely different data points. For the cat image, credit: Von.grzanka, CC BY-SA 3.0 https://creativecommons.org/licenses/by-sa/3.0, via Wikimedia Commons

- brightness and color manipulations

Some of the listed image transformations bring about the creation of empty areas: there are several approaches for filling these, from picking a background color, to duplicating nearby pixels in the original image.

Typically a deep learning library asks you to select what types of transformations you want to apply, and to specify some ranges of application. For example, a library can ask you what the maximum shift is, in pixels, that could be applied to your images. Then, during training, the exact amount of shift will be decided by picking a random value between zero and the upper limit that you defined.

A gallery of examples of image transformations for data augmentation can be found in the accompanying notebook in Sect. A.13.

Data Augmentation for Tabular Data, Sequences, Sounds...

As we said, data augmentation is by far more commonly applied to image data sets. However, it is possible to augment other types of data and reap the benefits of data augmentation. The only drawbacks are practical. First, it will really depend on the specific problem, so you will need to figure out what kind of transformations make sense for your data and what magnitudes and ranges should be considered. Second, and this is a consequence of the first point, you will usually need to implement a custom data transformation.[2]

For numerical data (tabular or time series) the easiest data augmentation technique is to add random noise. This builds on the observation that all experimental

[2] In Python implementations this usually takes the form of a *lambda* function, a small anonymous function that is called as part of the data preprocessing.

data are noisy, and repeated measures on exactly the same samples will produce slightly different numerical values anyway. A common choice is to sample the noise from $N(0, \sigma)$ that is, a Gaussian distribution with mean equal to zero and some variance of your choice. The exact value of σ commands the intensity of the added noise and thus how much the augmented data will depart from the original input. This may be regarded as an additional hyperparameter (yet another!) of DNN models to be fine-tuned through cross-validation. Having mean equal to zero avoids introducing unnecessary biases.

For sound data it is possible to apply audio-specific transformations like:

- time stretching: changing the speed of the audio without changing the pitch
- pitch shifting: changing the pitch of the audio without changing the speed
- volume adjustment: adjusting the volume of the audio
- time shifting: shifting the audio forward or backward in time
- adding noise: adding different types of noise to the audio, such as white noise[3] or background noise

Other types of data (e.g. 3D images) would require problem-specific solutions. The general rule is to try and separate the actual piece of information from details linked to the specific measurement, and then transform the data so to simulate other measurement conditions. Remember that data augmentation works by creating a diverse set of examples that help the model to generalize better. The limits of how much data can be augmented will be explored with the usual cross-validation framework.

6.3 Image Segmentation

In its simplest version, image segmentation[4] is the task of separating the subject from the background. It can be considered a classification problem, but the classification is repeated for each pixel. For standard classification problems we need a training set where each sample is labelled with its ground truth. For segmentation we need the ground truth to be expressed at the pixel level and collected in *masks*. A mask is another image, of the same size as the original one that we are trying to segment, in which each pixel contains the encoding for one or more classes. An example of a truth mask is the following, where each pixel can belong to either one of two classes, cat or no-cat (Fig. 6.2).

[3] "White noise" is a randomly generated stationary (no trend) signal series with zero autocorrelation.

[4] We refer to *image* segmentation because it is by far the most common application of segmentation and it is easier to make meaningful examples using images. However, it is straightforward to adapt the core concepts to other types of data, i.e. sound segmentation would be the task of extracting pieces of important sound, such as a human voice pronouncing "OK Google", from the noisy soundscape of an ambient recording.

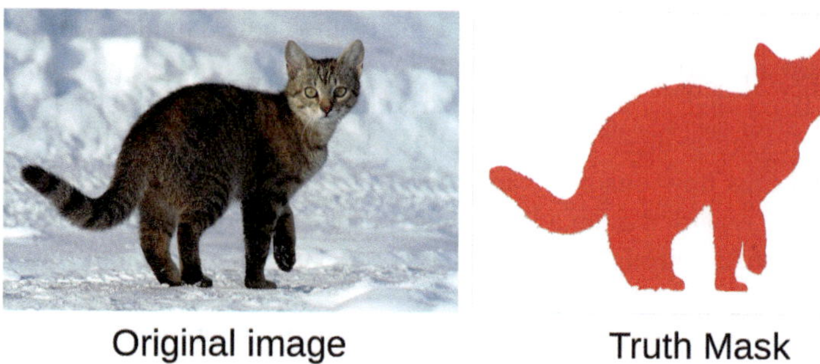

Fig. 6.2 An example of a truth mask for data segmentation. Each pixel is labelled, in this case either as cat (red) or background (white). Given that there are only two classes in the example, doing image segmentation becomes equivalent to doing binary classification on each pixel. For the cat image, credit: Von.grzanka, CC BY-SA 3.0 https://creativecommons.org/licenses/by-sa/3.0, via Wikimedia Commons

While image segmentation is conceptually very similar to standard classification tasks, its peculiar features command some differences. First, we will need to adapt the loss functions. Second, we will need to adapt the neural network architectures.

Loss Functions

As we recall from Sect. 4.3, a loss function is a measure of the errors the network makes when processing a sample. With simple classification the loss function is straightforward: if the sample is predicted by the neural network to belong to the correct class, it will obtain a good score (low loss value); otherwise it will get a bad score (high loss value).

While image segmentation can be considered as repeated pixel-by-pixel classification, we still need a single loss function, meaning that an image will result in one number after the network has processed it (forward propagation at any given epoch).

In this section, we will see three different loss functions that can be used for image segmentation: pixel accuracy, intersection over union (IoU) and the Dice coefficient. The first loss function for image segmentation that we are going to examine is *pixel accuracy*: it's the most natural solution, and it is easy to understand, but it is also a terrible choice, given the inherent flaws that make it unusable for any sensible application. The study of the mechanics of pixel accuracy, however, will give us the opportunity to understand some important specific aspects of image segmentation that are unique to this type of problem. The other two loss functions, *IoU* and *Dice coefficient* are only slightly more complex, and are commonly used in real world segmentation applications.

Pixel Accuracy

This is the simplest loss function for image segmentation and is defined as:

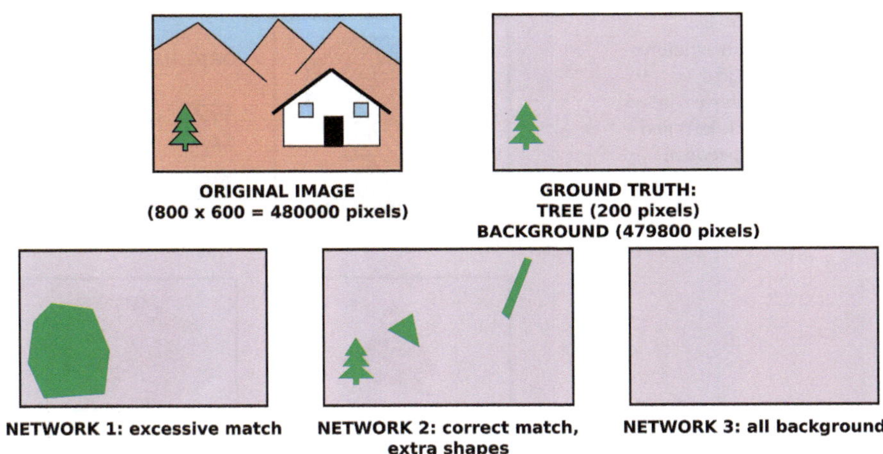

Fig. 6.3 An example of why pixel accuracy is a poor metric for image segmentation. In the top row we see a sample image with its ground truth mask. In the bottom row we see the output of three different networks. The first two are imperfect and find "tree pixels" even where there are none. The third one is stuck to the majority class and classifies all pixels as background

$$\text{Pixel Accuracy} = \frac{\text{Number of pixels correctly classfied}}{\text{Total number of pixels in the image}}$$

Pixel accuracy is thus a number between zero (no pixel correctly classified) and one (all pixels correctly classified). It is easy to understand since it's easily translated to a percentage. Indeed, we can easily think that we have a clear idea of the results if we are told that a network classifies well, e.g. 30% or 80% of the pixels. Unfortunately it is easy to see a limitation of this metric. Consider the image in Fig. 6.3.

In this example we are interested in finding all pixels containing a tree, as highlighted by the ground truth mask. The image is small by modern standards, only 800 pixels wide by 600 pixels tall, for a total of a little less than half a million pixels. Our area of interest is the tree on the left, covering about 400 pixels. We can already guess the issue: the area of interest is thus $400/480\,000 \simeq 0.083\%$ of the total number of pixels. In the field of standard classification problems this would be a very unbalanced data set. As such, it is bound to suffer from one flaw: Network #3, which classifies all pixels as background (i.e. the majority class), reaches an extremely high accuracy, so high that it may be preferred to other networks that, albeit not perfect, are at least trying to actually solve the problem at hand.

This case is very common, since the object of interest often occupies only a small fraction of the total image—we, as humans, may have a different perception, since our brains are used to focussing on the details and give them a lot of importance, but the raw data tell a different story.

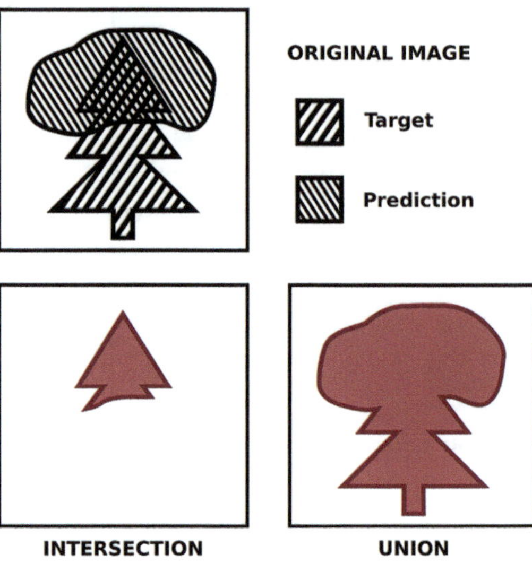

Fig. 6.4 Basic image segmentation nomenclature: prediction target and predicted area (above); below, the intersection (I: left) and union (U: right) between target and prediction

IoU (Intersection over Union), a.k.a. Jaccard Index

For the description of **IoU**, let's first introduce the following nomenclature for (i) the prediction target, and (ii) the actual DNN prediction: the Boolean algebra operations of **intersection** and **union** between the two sets "target" and "prediction" (Fig. 6.4).

We can then introduce the Jaccard index, colloquially known as IoU (intersection over union), which is a metric for image segmentation defined as:

$$J(y, \hat{y}) = \frac{I}{U} = \frac{|\text{Target} \cap \text{Prediction}|}{|\text{Target} \cup \text{Prediction}|}$$
$$= \frac{|\text{Target} \cap \text{Prediction}|}{|\text{Target}| + |\text{Prediction}| - |\text{Target} \cap \text{Prediction}|}$$

where y is the truth mask, $\hat{y}$ is the predicted segmentation, vertical pipes $||$ represent the operation of counting the number of pixels in the area described between them, and $\cap$ and $\cup$ are the mathematical symbols for intersection and union, respectively.

Following the definition, if the segmentation is perfect (all target pixels are identified as such) the intersection will completely overlap with the union, meaning that the Jaccard index will be 1. As soon as we move away from the perfect case, the union, by definition, will cover a larger area than the intersection, and thus the Jaccard index value will get progressively lower. In the extreme case of completely botching the segmentation there is no intersection and thus the index is zero.

When used as loss function, IoU would be needed to display a different behaviour, i.e. the opposite pattern, since low values (ideally: zero) are expected to represent good performances. Loss, in fact, is akin to measuring an error, so that fewer errors mean better segmentation. The Jaccard index is thus transformed into the Jaccard distance:

$$d_J(y, \hat{y}) = 1 - J(y, \hat{y}) = \frac{U - I}{U}$$

By definition, $d_J(y, \hat{y})$ is zero for a perfect segmentation and grows progressively, up to a maximum of one, for less than perfect predictive performances. In this way, IoU can be used as loss function. Note that this is often done transparently by deep learning software libraries, so that they may tell you that they are using the Jaccard index as loss, while internally they are actually using the Jaccard distance.

Dice Coefficient

More correctly called the Sørensen–Dice coefficient (SDC), but commonly know just as the Dice coefficient, this segmentation metric is defined as:

$$SDC(y, \hat{y}) = \frac{2\,|\text{Target} \cap \text{Prediction}|}{|\text{Target}| + |\text{Prediction}|} = \frac{2I}{U + I}$$

This metric reaches the maximum value of $SDC = 1$ only for perfect segmentation, i.e. where target and prediction completely overlap, and thus also coincide with their intersection. Worse segmentation will shrink the numerator but leave the denominator untouched, and thus it will lower SDC until it reaches zero for a completely botched segmentation.

As with the Jaccard index, SDC too can (and should) be transformed into a distance before it can be used as a loss function:

$$d_{SDC}(y, \hat{y}) = 1 - SDC(y, \hat{y}) = \frac{U - I}{U + I}$$

An Example Network Architecture for Segmentation: U-net

Let's consider the shape of a network designed for image segmentation. Inputs are going to be images, and we know that convolutional neural networks are very efficient for image processing. The standard topology for this kind of network, as we have seen in Sect. 4.6, vaguely resembles a funnel. Input data start wide (in terms of height and width) and thin (one channel for a monochrome image, three for RGB). Proceeding through the network the data shape changes, its face becoming smaller but also thicker since more and more channels are added. At some point, the convolutional part ends and a fully connected part begins.

This generates a data representation that is very compact, since the data shape shrank along two dimensions (height and width) and grew only along one (thickness). It's a good starting point, but it is not usable for the task of image segmentation, since the required output is a prediction of a truth mask. In other words, we need the network to output the same data shape that has been given to it as input.

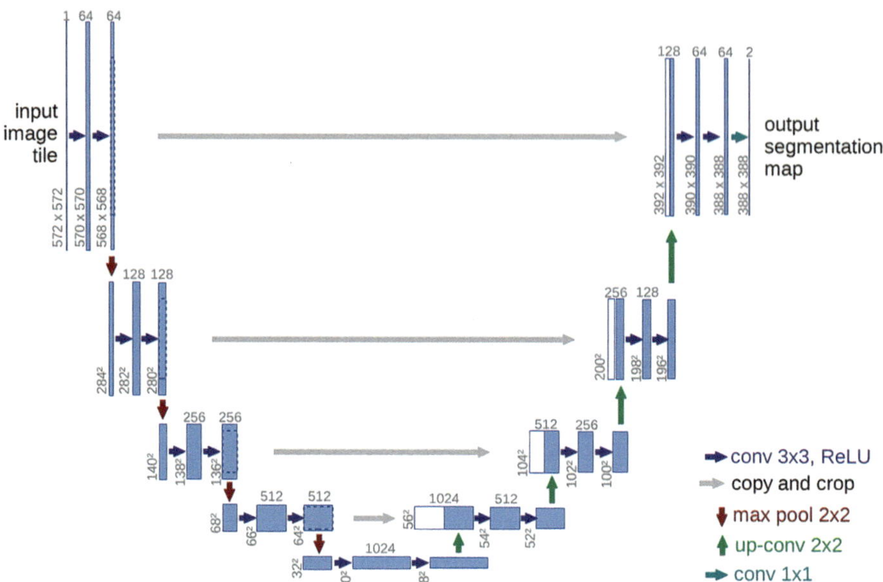

Fig. 6.5 U-net architecture (example for 32×32 pixels in the lowest resolution). Each blue box corresponds to a multi-channel feature map. The number of channels is denoted on top of the box. The x-y-size is provided at the lower left edge of the box. White boxes represent copied feature maps. The arrows denote the different operations

The most intuitive solution at this point would be to do the same thing that created the very compact data representation, but in reverse. In other words, we need to upscale. And in fact that is exactly the solution pioneered by Olaf Ronneberger, Philipp Fischer, and Thomas Brox in their seminal paper of 2015 [99]. They called their network U-Net, and it looked like that in Fig. 6.5.

On the left branch there is what is more or less a standard neural network architecture with convolutional and pooling layers. The data shape starts as a 572×572 image with only one channel, and through the network it is progressively shrunk and thickened so as to reach the size of a 30×30 cuboid with 1024 channels that you can see at the bottom. Then (almost) exactly the same architecture is run in reverse and the data is upscaled to output a 388×388 one channel image that contains the predictions for the segmentation mask.

Commonly, the downscaling branch of the network is called the "encoder", while the upscaling branch is called the "decoder". Please note that this nomenclature is posthumous to the original paper. Other than the two branches forming the "U" shape (and thus giving the architecture its name), you may notice the presence of *skip connections* joining earlier and later stages of the process. These skip connections allow the decoder to access the feature maps produced by the encoder, which helps to preserve fine details and information from the encoder while performing the upsampling in the decoder.

The U-Net architecture is well suited for image segmentation because it can effectively localize objects within an image while also preserving fine details and high-resolution information. While today somewhat dated, U-Net was groundbreaking when it came out, improving on many at-the-time state-of-the-art records for competitions on image segmentation, such as the ISBI cell tracking challenge 2015.

Today more complex architectures are available, like SAM (Segment Anything Model [100]), which internally leverages large language models and is trained on a very extensive dataset. Some general architectural features of U-Net remain, however, because they are intrinsic to the segmentation task. For example, all segmentation networks, whatever the internal structure, need to output a big matrix of classified pixels, and their performance is going to be evaluated using the same metrics (and loss functions) that we have discussed.

Stepping up the Difficulty: Instance Segmentation

So far we have talked about what is called *semantic* image segmentation. In simple terms, we wanted the network to "just" be able to classify each pixel of an image, this being background, a tree or a tumor cell. We can go further and ask the network to separate the different instances of the classified objects. More formally, the goal of instance image segmentation is to identify and distinguish between individual objects in an image, even if the objects belong to the same semantic class (Fig. 6.6).

This is a harder task. A training dataset for instance image segmentation typically consists of images with annotated instances. This means that each instance of an object in the image has been manually segmented and labeled. So the training data should be representative not only of the variations in object appearance and scale that will normally be encountered in the real world, but also of the ways in which the object instances can interact with each other. Often borders are ill defined, and it can be very difficult—even for humans—to establish exactly where one target object (a cell, a plant, a person) ends and another begins.

An example of a successful network for instance segmentation is Mask R-CNN [101]. This architecture is interesting because it introduces an element that we don't often see in networks: a branch. Mask R-CNN is built on a previously existing network, Faster R-CNN [102], designed for object recognition.[5] Mask R-CNN "simply" adds a parallel branch to Faster R-CNN so that the pixels *of the bounding box* are classified as either being part of the target object or being background. This is the key concept of this architecture, and the main reason why it has been so successful: the hard problem of instance segmentation is split in two simpler tasks, namely object detection (already solved by Faster R-CNN) and binary semantic

[5] *Object recognition* is a task conceptually similar to instance segmentation, since it still requires locating all instances of the target object in the image. But instead of a pixel-by-pixel classification, the output of object recognition is just a bounding box, i.e. the x-y coordinates of the top-left and the bottom-right corners of a rectangle that encompasses the object instance (see Sect. 6.4).

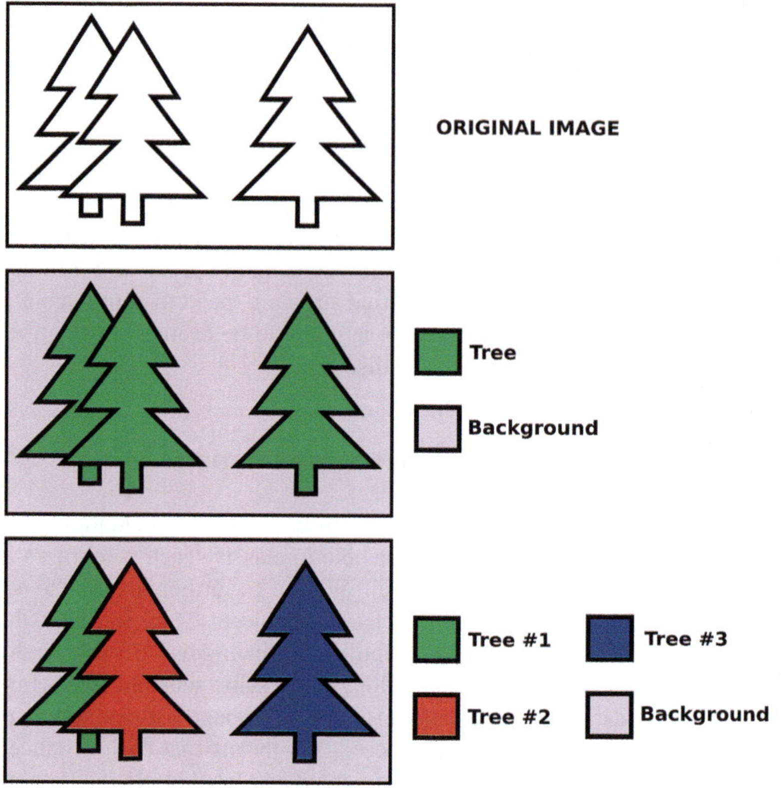

Fig. 6.6 The difference between instance (bottom) and semantic (middle) image segmentation

segmentation of only one subregion of the image that contains one instance of the target object.

6.4 Going Beyond Classification

Throughout this book, we have mainly described DNN models—despite the different architectures employed—developed to tackle classification problems (binary or multinomial) starting from a variety of input data types (tabular data, images, sequences). Deep learning is however very versatile, and neural networks can obviously be used to solve many additional types of statistical problems, some of which we have already mentioned briefly here and there (e.g. regression, pattern recognition, text generation).

In this section we try to give an overview (far from systematic) of the main types of problems for which DNNs are successfully used, reminding the reader that there are many new areas in which the capabilities of deep learning keep on being tested, and there is virtually no limit to the creativity of the scientists involved in this epic intellectual activity. Many of the fields we mention re-use—oftentimes in a very creative way—concepts and DNN architectures that we have already described in this book; some introduce new concepts and advanced architectures that are very powerful and open up new avenues of research and technological applications.

- **Regression problems.** In Sect. 2.5 we discussed the difference between classification and regression problems: in a nutshell, what changes is the target variable, which is not a category (class: ordered or not) to guess, but a continuous number in $[-\infty, \infty]$ (outcome) to predict. With respect to basically all other aspects (preprocessing, network architecture, hyperparameter tuning etc.) the DNN model remains unaltered, with the exception of the loss function (e.g. MSE) and of the metrics used for model evaluation. For a practical implementation and examples see the notebook in Sect. A.9.
- **Object detection (OD).** This is an extension of image processing and segmentation, as it reads an image and identifies and locates multiple objects within it (labelling them). Image segmentation groups together pixels with similar properties; in OD these groups of pixels are then localised within the image and labelled. OD is a central technology in **computer vision**[6] applications, based on network architectures like RCNN (region-based CNN), YOLO (you only look once), SSD (single-shot multibox detector) and RetinaNet. More recently, architectures based on transformer models borrowed from sequence analysis, are gaining popularity: examples include DETR (detection transformer [103]) and Co-DETR. Key challenges include scale variations, partial occlusions (the object may be partially blocked by other overlapping foreground objects), background contrast (lack of), appearance variability (e.g. rotation, lighting).
- **Pattern recognition (PR).** Pattern recognition is somewhat similar to object detection, but extended to multiple types of data. The core to PR is finding regularities—patterns—in the data to then be used for a variety of purposes (predictions, clustering, decision-making, content generation etc.). Applications include OCT (optical character recognition), the analysis of time series data like stock market prices, reading barcodes and QR-codes, the detection of functional motifs in protein [104] or DNA [105] sequences, and even emotion recognition [106].
- **Text processing and machine translation.** The analysis of text data and, more in general, of human language, is one of the areas of deep learning that has produced the most spectacular results. Text can be read and given a score of "positivity" or "negativity", like in the **sentiment analysis** of e.g. place or product reviews, comments under blog posts, feedback from users; another example is **spam detection** where the content of emails is read and classified to identify useful

[6] Computer vision refers to the ability of DNN models to read and understand image data (static images or videos), in a way similar to what humans—and other animals—do when they look at reality with their eyes.

messages and distinguish them from spam email. A relatively more advanced problem is that of **text summarization**, in which the neural network reads and "understands" the text, and then produces a meaningful concise summary. Text summarization can be broadly subdivided in **Extractive Summarization** (important words and phrases are extracted) and **Abstractive Summarization** (new phrases are generated that capture the meaning of the input text) (more details in the review [107]). The analysis of text data typically involves several steps: first, the text is cleaned, e.g. by removing common uninformative words (**stop words**, like *and* or *the*), punctuation, capitalization. After cleaning, the text is **lemmatized**, i.e. words are reduced to their base form or "lemma" (e.g. "takes" and "taking" are both converted to "take"), and **tokenized**, i.e. sentences are split into individual words. The next step is **vectorization**: since neural networks only take numbers as input, words must be converted from strings to a numeric representation; vectorization consists of transforming sentences into large sparse matrices of 0s and 1s based on the **dictionary** of lemmatized and tokenized words. Finally, the text data is **embedded**, a fancy word meaning that the large sparse matrix resulting from vectorization is condensed into a smaller dense numeric matrix that somehow captures the meaning of the words (more details in Sect. 5.4 ("Embeddings: a new representation for sequence data"), in [108] and in the accompanying notebook in Sect. A.14). A popular and quite remarkable application is **machine translation**, which is a specialised type of deep learning that can automatically translate text from one language to another. Machine translation is typically based on network architectures that involve an **encoder** that reads and encodes the input text in language "A", and a **decoder** that converts the encoded input to language "B". Machine translation models are now capable of achieving excellent performances [109] and are used for very reliable applications.[7]

- **Sound recognition.** Neural networks can process not only written text, but also spoken text and—more generally—any sound data: sounds are essentially sequence data (Sect. 5.1), and are one of the structured types of data that are efficiently and effectively processed by DNN models (Example 5.1). Neural networks that analyse sound data are used for applications like voice recognition on smartphones, the classification of songs and bird sound detection [110]. Sophisticated applications of DNN models for the analysis of sound data involve the identification of abnormal respiratory sounds in human medicine [111], or the detection of respiratory disease outbreaks in veterinary medicine/epidemiology [112]: these are both examples of pattern recognition applied to sound data. Importantly, neural networks are not limited to the analysis of audible sounds, but naturally extend to infrasounds (frequency < 20 Hz) and ultrasounds (frequency $> 20{,}000$ Hz).

[7] Google Translate: https://www.deepl.com/en/translator; DeepL Translator https://www.deepl.com/en/translator.

- **Content generation.** One of the most spectacular—or at least most trendy and popular—examples of the wonders of deep learning is provided by the recent explosion of generative **large language models** (LLMs), like those implemented in products like OpenAI chat-GPT, Google Gemini, Microsoft Bing, etc. One of the engines of these models are GAN (generative adversarial network) models, where one network generates the content (e.g. a text), and the other network tries to differentiate the machine-generated content from the human-generated content (the "truth"). GAN can thus generate high-quality content that is difficult to distinguish from non-AI generated content [113]. Besides text content, image and sounds are also popular applications for generative DNN models.[8]

6.5 Transfer Learning

When reading about successful deep learning projects, it is very easy to be intimidated by the sheer size of the training data sets used. It is now a common occurrence to find networks trained on millions of images, and with text-oriented data sets the situation is even worse.[9] If you are approaching the topic of neural networks, and especially if you are a researcher trying to analyze your experimental data, chances are you do not have that kind of data available. The good news is that you probably don't need it.

Transfer learning is a technique that involves using knowledge gained while solving one problem to improve performance on a different, but related problem. In transfer learning, you have a (usually large) model trained on a (usually large) dataset for a specific task. The model is then used as a starting point for a new task. In simpler terms, you take an already trained network and adapt it to your problem.

This technique works because a big part of "learning" involves acquiring some very basic, very general concepts. If you are training a network on images, at some early stage your nodes will need to discover the concepts of line, angle, border, shade, and so forth. If you are training a network on sounds, the building blocks will be pauses, pitch and volume. Other applications will have their own basic components. Understanding and manipulating information at this core level requires time and effort but, luckily, it is quite a general task. So general that it can be easier and more efficient to take a network already trained to recognize the breed of a dog and adapt it to identifying tumors rather than starting from scratch.

To apply transfer learning successfully, we should keep in mind that there is a semantic progression through the network layers—that's the key point of *deep*

[8] SoundRaw https://soundraw.io/, Loudly https://www.loudly.com/ai-music-generator, DALLE-2 https://openai.com/index/dall-e-2/, Midjourney https://www.midjourney.com/home and many more.

[9] For reference, the version of ChatGPT released in November 2022 was trained on about 500 billion "tokens" of text data. For a more tangible reference, the entirety of Wikipedia was incorporated into the training dataset, and that amounted to only 3% of the total data considered.

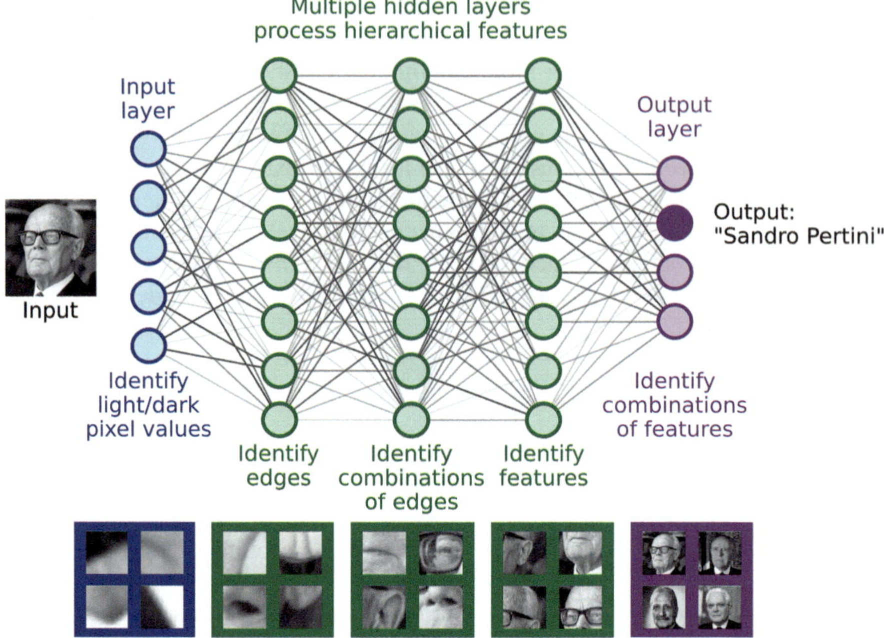

Fig. 6.7 Information is treated at progressively higher levels of abstraction through the network layers. In this example, a network is trained to recognize the presidents of the Italian Republic

learning. The first layers are usually about very fine details, the kind of minutiae that would make little sense taken out of context (e.g. light/dark pixels, as shown in Fig. 6.7). The next layers aggregate the information to higher levels of abstraction, to the point that the final layers are able to manipulate concepts like cars, species of birds, or sea currents.

To implement transfer learning, you need to follow these steps:

1. Acquire a big, already trained network. Many deep learning libraries already come with prepacked models, or you can download one from open-access publications or by any other suitable means
2. Remove the very last layer of the network and swap it with one that matches your precise problem so that the output is as expected (e.g. the network outputs the correct number of classes)
3. Freeze the majority of the network. Here, "freezing" means to lock the nodes so that no learning happens. It's a very common functionality that comes with modern deep learning libraries. Remember, you usually don't have enough data to train a very big network and, moreover, you are trying to leverage one that is already trained. You want to be sure to train the very last layer that you just added in the previous step, and maybe a few layers just before that one—the exact amount depending on your problem and data availability

4. Train the not-frozen layers on your data, check the performances

Transfer learning is particularly useful when considering very, very big networks. For example, Large Language Models (LLM) are trained on enormous datasets, and each training round costs millions of euros (or dollars, at that scale it doesn't make any difference). Luckily, as the deep learning frameworks become more and more standardized, it is pretty simple to import external models. For example, Keras comes with built-in models called "applications", which can be imported with a single command line [114]. Similarly, the Kaggle online repository [115] contains many models—some open sourced, some provided by companies like Meta under license—that can be imported into your project with little effort.

To see an example of implementation of transfer learning, please refer to the notebook in Sect. A.15.

Fine Tuning

Fine tuning is a trick, and you should consider it just as that. However, it may improve the network's performances a little, so it may be worth trying. In general, it is applied after transfer learning has been completed, i.e. when you have just finished training the final layer(s) of a large, pre-trained network. Here, the assumption is that you don't have enough data to train the complete network from scratch, i.e. if you unfreeze all the layers and train the model, performances will rapidly degrade.

However, the following procedure has proven capable of further improving a network's performances:

1. unfreeze the whole network
2. chose a *very small* learning rate, e.g. one tenth (or less) of what you used during transfer learning
3. do another round of training (on the whole network)

The idea is that, with a very small learning rate, the network will not undergo radical changes, but will simply look for improvements to the local optima (even if you do not have lots of data). Optimization is often compared to finding the top of a mountain. In this framework, fine tuning is to be considered as having already arrived at the top, and looking for the exact highest stone to climb onto to gain some extra centimeters.

The same concept can be applied to regular learning, that is, even when you are not doing transfer learning. The core idea is very similar to what is called "Learning rate decay", which is exactly about using smaller and smaller learning rates once the training is getting closer and closer to a local optimum (you don't want to jump away from there, but you do want to continue exploring the area for the "lowest spot").

6.6 Recipe for a Good Project

When you are approaching a new deep learning project, a good initial recommendation is to follow these steps:

1. Define a single metric
2. Define a target performance
3. Prepare your data/1: preprocessing
4. Prepare your data/2: split the dataset.
5. Do good on the training set
6. Do good on validation/test sets
7. Error analysis
8. As a last resort...

Let's inspect each one of these steps in order.

Define a Single Metric

Sometimes the target metric is given and there's little to discuss. Many other times, however, problems are more complex and multiple metrics can (and maybe should) be considered as viable descriptors of the model's performance. Unfortunately, the optimization algorithms will do their best to improve a single performance metric—a single number—and the choice of what metric to optimize can have strong consequences on the final behaviour of the model.

In Sect. 3.1 we listed the most common metrics available for classification and regression problems; additionally, other special cases relative to segmentation are presented in Sect. 6.3.

As stated, you can only optimize a single metric.[10] However, it is often possible to take into account several metrics at once, introducing the distinction between "optimizing" and "satisficing" metrics. Consider two models, a fast but rough one and a slow but precise one. Let's say that we are performing classification and the first model reaches a 10% error rate requiring 10 seconds to run. The second model reaches a 3% error rate, but requires a full minute. You can easily see that, depending on the final application, you may end up preferring either the first or the second model. As an example, any real time application will impose some constraints on the maximum execution time. This is a classical example of a *satisficing* metric: while it is nice to have a very fast model, the really important requirement is to not overshoot the allotted time slice.

[10] That's not completely true, and in fact there does exist an entire branch of optimization theory dedicated to *multi-objective optimization*, with important applications in economics, finance and engineering. For a gentle introduction on the topic, see [116]. For reviews of available methods, see [117] and [118].

If the total performance of your model can be described using N metrics, you may want to select one as *optimizing*, i.e. the one that will be used to drive the training. The other $N - 1$ metrics will then be considered as *satisficing*, i.e. each will have a threshold that needs to be respected. In this way, the comparison between models becomes a two-step process. First, discard all models that don't satisfy the constraints. Then, select the best model based on the optimizing metric.[11]

Define a Target Performance

What do you expect from your model? What would good performances look like? What about very, very good performances?

Given a problem, there does exist an intrinsic value called the *Bayes error rate*, which defines the best possible performance (i.e. the lowest possible error rate) for a classifier on that specific problem.[12] In simple terms, no classifier could ever do better than the Bayes error rate, which is truly the limit of optimality given a specific problem.

We are interested in knowing the Bayes error rate because it would give us an immediate reference point from which to evaluate the performances of our model. In practical terms, a network producing a 10% error rate looks very different if the Bayes error rate is either 1% or 9.5%.

Unfortunately, it is usually impossible to know the exact value of the Bayes error. The common approach is thus to look for proxies, and the best proxy is usually a panel of human experts. Let's say that you want to classify cells in a piece of microscope photography. You gather together five-ten microscopists, ask them to classify ten thousand images and then average the results. The final error rate would be a good proxy for the Bayes error rate, or at least for the human-based limits. You can easily adapt this approach to your problem and imagine a panel of experts classifying bird calls, migratory patterns, disease susceptibility, and so forth. The only problem with this approach is that it's very expensive, both in terms of money and time (experts tend to be busy people). It is, however, not impossible. If your problem is somewhat well known, you may be in luck: you may find that a panel already gathered years ago, and you can thus piggyback on other people's work. Try to stand on those giants' shoulders.

A second, more modest approach is to scan the state of the art for published classifiers and look at how they fare on your problem. This is cheaper and has the advantage that performances are often already in a directly comparable format. It is also a worse proxy for the Bayes error rate, since algorithms tend to come in

[11] Some readers may recognise here a certain similarity to **Linear programming** (**LP**, also known as linear optimization), in which you have an objective function to be maximised (or minimised) subject to multiple constraints.

[12] A similar concept for regression problems is the *irreducible error*. In the text we will continue referring to the Bayes error rate for simplicity, but the same reasoning applies to regression.

waves and improvement is not linear—as deep neural networks have demonstrated. It has its merits, however, since your direct competitors in any problem are the actual implemented solutions, not the theoretical limits.

Finally, if you really cannot find anybody who has ever faced your problem, you are in luck, since anything you produce will be the best solution that humanity has yet produced on the topic.

Prepare Your Data/1: Preprocessing

Cleaner data organization means less pain during training. The first aspect you need to consider is data preprocessing, since your data rarely arrives in a usable form. We discussed data normalization and augmentation in Sects. 6.1 and 6.2. Once you know what kind of preprocessing you are going to need, it is important to decide if you are doing to do it "on the fly" (i.e. as part of the model you are building) or once and for all before even approaching the neural network. This choice should reflect the final application you are building. Considering the preprocessing step as part of the application allows for further optimization, but at the price of slowing the training process, since there's more computation required. Focusing on just the network allows for a more agile development, but you should always consider that in many cases, somebody, somewhere, will need to take your raw data as input and transform it into a usable form.

Prepare Your Data/2: Split the Sets

At minimum, you'll need two sets of data: the training and the validation sets. You may need a third one, the test set, too. If you are lucky and have "enough" data, the split can be done once, statically, before starting any kind of model optimization. In this case, you can just split the data into big chunks that reflect the presence of strata: if your general dataset is comprised of 80% samples of type A and 20% of type B, both the training and the validation sets should have approximately the same sample-type ratio. Here, "type" is used loosely: it could (and should) represent the actual classes, but any additional meaningful subgroups present in your data should also be considered.

In practical terms, this kind of data-splitting is often implemented just by saving the training, validation and test data in different folders, so much so that the major deep learning libraries all have functionalities to import data automatically, if these are saved and stored in the expected folder structure.

A slightly more complex approach is required if you don't have "enough" data to do a static split and instead need a cross-validation. Unfortunately, there's no fast way to determine if this is the case, but a rule of thumb is that, if your sample count is in the hundreds, you should at least consider cross-validating your model. Cross-validation can be tricky to implement, because the split in folds should reflect the data stratification, plus there's more explicit control about loading data into memory, which can easily become a bottleneck (more details on cross-validation have been discussed in Sect. 3.2).

Do Good on the Training Set

It may sound obvious but, if your model doesn't work well on the training set, it will never make it on the validation set. Performances on the training set are always inflated, so you should expect very high accuracies/low error rates on the training set, at least in the first iterations.

If you reach average performances at best, it's time to double check what you have done so far. It may be due to some implementation errors, something wrong with your specific dataset, with the preprocessing, with the values of the hyperparameters, or maybe you do not really understand the problem. It's ok, it happens quite often and to everybody.

With reference to neural networks, remember that (very roughly) deeper means smarter. When experimenting with network topologies, a good bet for improving the performance on the training set is to add layers.

Do Good on Validation/Test Sets

This is the hard part. If you reach this stage, you usually have a network that performs very well on the training set, maybe reaching 99% accuracy or other stellar, improbable values on your metric of choice. Then, you measure the same thing on the validation set and obtain terrible scores. The main culprit here is overfitting. You overshoot the network, which is now too big for the available data. Here's a list of the stuff you should try to optimize, in decreasing order of importance and/or simplicity of implementation:

1. learning rate: too big and your model jumps around the feature landscape, too small and it takes forever to converge. You may also want to use a scheduler to progressively reduce it during the training cycles
2. reshape the network: by removing layers and, in the second instance, nodes in some/all of the existing layers
3. change optimizer: several options come with the major libraries, with ADAM [119], RMSprop [120] and Stochastic Gradient Descent (SGD) [121] being the most common choices. See Sect. 4.7 for more details.
4. use anti-overfitting techniques like L1/L2 regularization, dropout and early stopping. See Sect. 4.8 for more details.

Improving the performances on the validation set may come at the cost of worsening performances on the training set. This is fine, and there's no need to panic. You will have eliminated all the overfitting once the performances on the training and validation sets are similar.

Table 6.2 Confusion matrix from: Model 1

	Predicted as A	Predicted as B
Samples of class A	35	8
Samples of class B	10	42

Table 6.3 Confusion matrix from: Model 2

	Predicted as A	Predicted as B
Samples of class A	27	16
Samples of class B	2	50

Error Analysis

This step is optional and often overlooked, but it may be crucial for at least understanding *why* your model is not doing well. Imagine having two models and, as a consequence, two confusion matrices. The first model results in the matrix shown in Table 6.2. While the second one gives that in Table 6.3.

While the total number of errors remains the same, it is clear that the first model is more or less symmetrical towards the two classes, while the second one is not. In fact, the second model "prefers" class B, and it's more inclined to classify a sample as B rather than A.

This is the core of error analysis for classification: you need to examine how your errors are distributed, and ensure that your model is equidistant from all classes. In the case of regression problems, the analysis is similar in principle, but you need to understand whether the model tends to under- or overshoot. If you discover that the model is biased, you can then intervene accordingly, e.g. on the loss function.

One possible outcome of error analysis is that you conclude that your problem is too difficult as it is formulated. If this is the case, it may be useful to split the problem into two or more simpler sub-tasks. For example, imagine a classifier that needs to separate samples belonging to the four classes A, B, C, and D. Error analysis may indicate that the last two classes are often mixed up. Instead of forcing a single four-class model, it may be better to have a first model classifying samples as A, B, or C+D, with the last group containing the two classes that are often confused. Now the problem is simpler and the performances will probably improve. Then, a second model will be designed and trained only on the specialized task of distinguishing samples from the classes C and D which, hopefully, will be a simpler task.

As a Last Resort...

Get more data.

6.7 Visualization of Deep Learning Models: A Primer

With most machine-learning methods it is possible to extract variable importance from prediction models: the input variables (features) can be ranked for their relevance with respect to the modelled target outcome, based on different metrics. Neural networks are often blamed for being "black boxes", i.e. phenomenal predictive machines that have the downside of not being interpretable: is this really so? For a brief illustration of the techniques used to shed light on the blackbox ("explainable AI"), see Sect. A.17.

6.8 Can We Do Feature Selection with Neural Networks?

This applies mainly to tabular data but, with a little ingenuity, it could be extended to other data types. Since the first layer of a network receives as input the actual numerical values of the features, it stands to reason that it should be able to discriminate between important and less important features. In fact, each node of the first layer *weighs* each input feature, and bigger weights (in absolute terms) somehow mean more importance. If this line of investigation works, we could then examine the behaviour of the second layer, and the third one, and so on, and maybe even extract some insights into not only the data, but also the network itself.

Luckily, we can easily investigate this hypothesis. Let's use the very well-known Wisconsin Breast Cancer Dataset [122], comprising 32 features on 569 samples for binary classification (either 'malignant' or 'benign').

This problem is relatively easy, and even very small networks can achieve satisfactory accuracies. But to investigate the capability of neural networks to actually discriminate useful features from useless ones, we can use a "sleight-of-hand" trick and add a new, completely random feature. Let's call it RANDOM_VALUES and sample its values from a random uniform distribution. A feature built in this way has no link whatsoever with the target class to be predicted. It should be easy for the network to discriminate. This feature is then added to the dataset and presented to the network during training.

Let's start with the simplest possible network: a single node with sigmoid activation function.[13] After a round of training, we open our single node and extract each weight assigned to all features. Since training is a stochastic process and we may have to split the data into training and validation sets, we repeat the whole training a few times, each time taking notes of the weights that the network assigns to each feature. Once all data is collected, we take the absolute value (our hypothesis is that big weights are important, whether they are positive or negative). We then rank the feature from the one that received the largest weight to the one that received the smallest weight (on average).

[13] In other words, we are actually doing logistic regression.

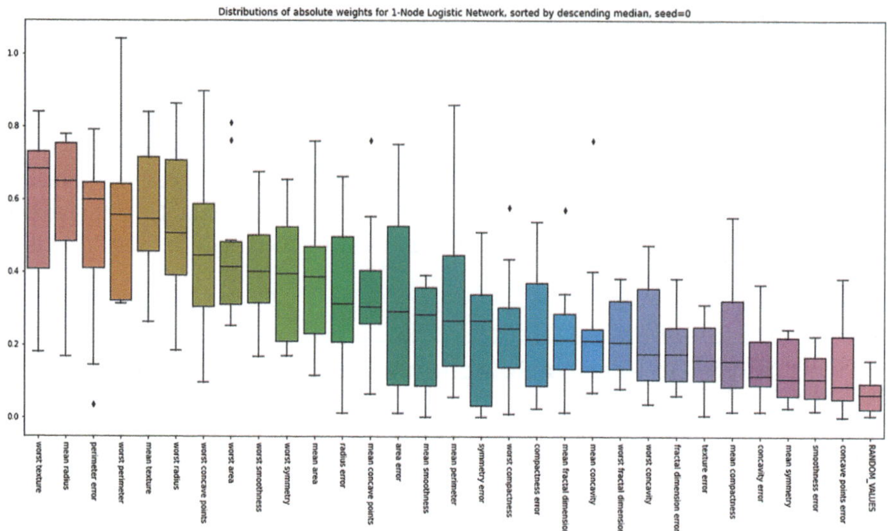

Fig. 6.8 Distribution of (absolute values of) weights assigned by a 1-node logistic network to the features of the Wisconsin Breast Cancer Dataset with the addition of the dud feature called RANDOM_VALUES

The results are in Fig. 6.8, and it seems a promising start: the network shows a clear preference among features, and our fake, random feature is ranked at the bottom (far right in the figure). It is true that the network we are using is very simple, with just one node and, in fact, using a network like this is a complicated way of just doing logistic regression. Nonetheless it seems that we are on an interesting track.

What will happen with more complex networks? Let's do the smallest possible step-up and redo the above analysis with a two-layer network: the first layer has 16 ReLU-activated nodes, the second layer has only one sigmoid-activated node (the output). In this experiment, each node of the first layer will have 33 weights (the 32 original features plus the extra random one). The only node of the output layer will have 16 weights (one per each node of the previous layer). We are ignoring the bias terms. The results are reported in Figs. 6.9 and 6.10.

Things are getting confusing. It is true that the dud feature RANDOM_VALUES still ranked last, but the distribution of its assigned weights is very, very similar to that of the other features. In fact, the weight distributions are much more homogeneous than what was obtained with only a single node, a sign that the discriminative power of the network is waning. Moreover, it's very hard to make sense of weights from the second layer. In fact, after the first layer the connection with the original features is completely broken, and there's no way to give the weights a sensible label—they are simply marked with a progressive number, from zero to fifteen.

This example shows a general trend: as the network grows, it becomes harder to actually understand its inner workings. Moreover, its "reasoning" becomes

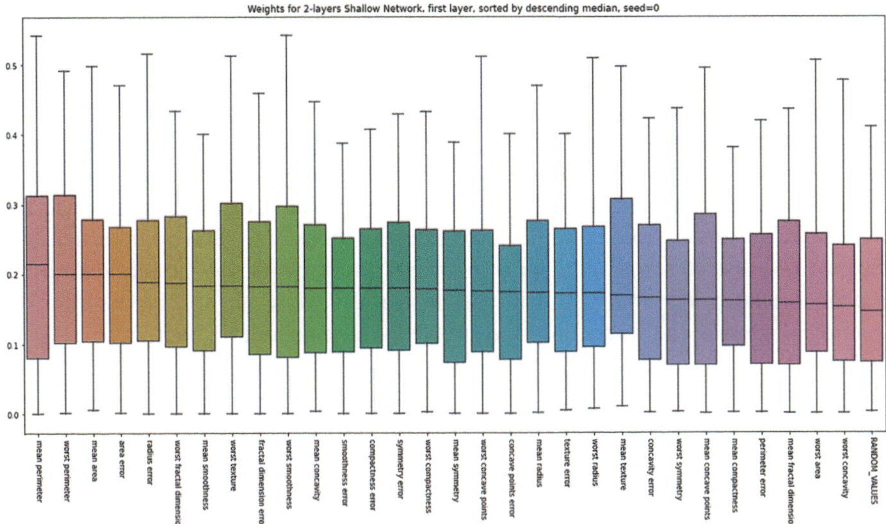

Fig. 6.9 Distribution of (absolute values of) weights assigned by the first 16 nodes of the first layer of a two-layers network to the features of the Wisconsin Breast Cancer Dataset with the addition of the dud feature called RANDOM_VALUES

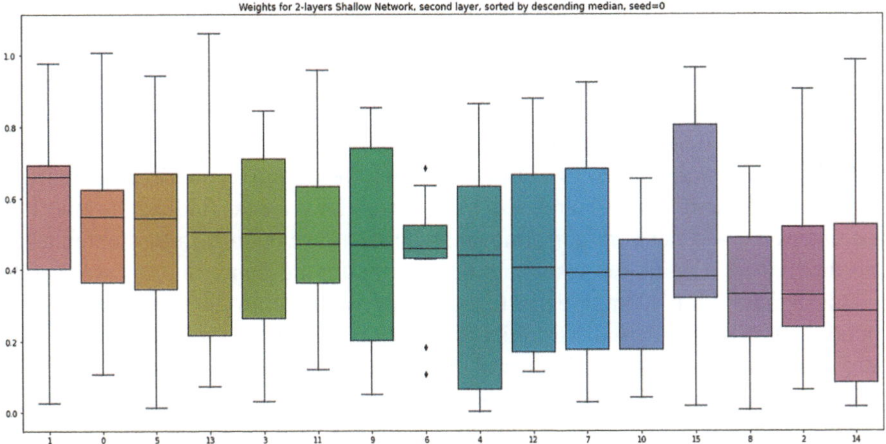

Fig. 6.10 Distribution of (absolute values of) weights assigned by the output node of a two-layers network to the features of the Wisconsin Breast Cancer Dataset with the addition of the dud feature called RANDOM_VALUES

confused, and it's hard to tell what the network considers important and what it doesn't. Some kind of intelligence is still there—the predictions are very good on this dataset—but it becomes impossible to grasp it.

More advanced techniques for feature visualization and network dissection belong to the expanding research field of **explainable deep learning**. These include methods that identify which nodes in the network are more active (e.g. have the larger activation values), or which features have the largest influence on the performance of the network. This can be obtained, e.g., by removing a feature and observing the change in the network performance, or by building a "white-box" model that mimics the neural network but is explainable, or even by engineering artificial adversarial input examples with extreme or absurd feature values that will more or less disturb (or not) the performance of the network. For an introduction to explainable deep learning, you can read this review work [123]. The code for the example discussed in this section is available in the notebook in Sect. A.16.

6.9 Can We Tune One Hyperparameter at a Time?

During deep learning training there are many hyperparameters to be tuned—learning rate, batch size, number of layers, number of units, activation functions ... to name just a few. Given that for each hyperparameter you are required to test a number of values, the final tally of combinations becomes very high, very quickly. The curious researcher might wonder whether we really, really need to try them all. Would a simpler approach work? Can we tune one hyperparameter, then move on to the second one, and so forth, until we have optimized each one and be happy with it?

The time savings would be staggering. Let's just look at a small numerical example. Let's say that we need to tune three hyperparameters, A, B and C. For each one we need to test ten values, $A_1, A_2, \ldots, A_{10}$, for the first hyperparameter, and similarly for the other two. Let's say that we optimize one hyperparameter at a time. We start from A and test its ten possible values, finding the one that guarantees the best network performance. We then move on to B (a further ten tests) and finally to C (again, ten tests) and declare the tuning done. In total, we had 30 rounds of calibration. Compare that to testing all combinations, which requires 1000 rounds ($10 \times 10 \times 10$), and our initial proposition (tuning one at a time, rather than combinations) becomes very interesting and tempting. Given that a single round of training can, for example, require an hour, there's a lot to be gained.

This approach, which may be called *greedy stepwise model selection* [1], has its own dignity and may be considered a good exploratory approach. It has a clear limitation, though: what if some of the combinations we did not try were better? The simplest counterexample can be constructed with just two hyperparameters A and B, each one having two possible values A_1, A_2, B_1 and, B_2. Let's say that our metric of interest is the error rate (i.e. lower is better). Consider the following fitness landscape (Table 6.4).

It is immediately apparent that (A_1, B_2) is the best configuration. However, when optimizing one hyperparameter at a time, it is possible to skip it completely. Consider the following implementation of *greedy stepwise model selection*, described in the algorithm below:

Table 6.4 A fitness landscape: each cell contains the fit for a neural network trained using a specific combination of hyperparameters

	B_1	B_2
A_1	10	0.1
A_2	5	1

```
1 best_solution = {A_1, B_1, C_1, D_1, ...} ;
2 best_fit = evaluate(current_solution);
  /* Optimizing for hyperparameter A                                  */
3 foreach A_i in {A_1, A_2, A_3, ...} do
4   | current_solution = {A_i, B_1, C_1, D_1, ...} ;
5   | current_fit = evaluate(current_solution);
6   | if current_fit < best_fit then
7   |   | best_solution = current_solution ;
8   |   | best_fit = current_fit;
  /* At this point we found A*, the best value for A keeping
     fixed B_1, C_1, D_1, ...                                         */
  /* Let's optimize for hyperparameter B                              */
9 foreach B_i in {B_1, B_2, B_3, ...} do
10  | current_solution = {A*, B_i, C_1, D_1, ...} ;
11  | current_fit = evaluate(current_solution);
12  | if current_fit < best_fit then
13  |   | best_solution = current_solution ;
14  |   | best_fit = current_fit;
  /* At this point we found B*, the best value for B keeping
     fixed A*, C_1, D_1, ...                                          */
  /* The same procedure applies to all other hyperparameters.
     */
```

This pseudocode (above algorithm) is a straightforward implementation of the discussed algorithm (please note the "i" footer in lines 4 and 10) and supports an arbitrary number of hyperparameters and their values. The key observation is that even when we are optimizing the very first hyperparameter (e.g. hyperparameter A) we *need* to assign some values to all the other ones (hyperparameters B, C, D, ...) otherwise either the network or the optimizer cannot work.

Given the above code, the example reported in Table 6.4 becomes clearer. We fix B to B_1 and optimize for A, finding the optimal value $A^* = A_2$, which becomes part of the final solution (in other words, after optimizing for A, we have completely excluded the first row of the table). We then optimize for B and find that our final solution is $(A^*, B^*) = (A_2, B_2)$, with a fit ten times worse than the true optimal one.

So, is this strategy really garbage? Not really. It has zero guarantees of finding the optimal solution, but it's a solid starting point. Given that it is unfeasible to test all possible configurations, a first round of greedy stepwise model selection can produce a reasonable solution, which can then be improved by other means (if time and resources are available). Keep in mind that, even though other more sophisticated techniques do exist, the general approach is similar: start from the current best solution and improve one (or more) hyperparameters on it.

For a more in-depth analysis of hyperparameter optimization with reference to neural network please refer to Google Brain's *Deep Learning Tuning Playbook* [124].

Part IV

Success and Failure Cases of Deep Learning Applications in Biology

7 Deep Learning Research and Applications: A Few Interesting Stories

The marriage of deep learning and life science has opened unprecedented avenues for exploration, from decoding the intricacies of genomics to unraveling the mysteries of cellular processes. At this crossroad, we find scientists who are actively improving the state of the art, either applying new techniques to old problems or completely inventing new approaches.

In this exclusive section, we bring to you insightful interviews with these forward-thinkers and provide a series of conversations that shed light on their research, as well as their personal paths. We will uncover the stories behind their groundbreaking research, the challenges they've faced, and the transformative impact of incorporating deep learning methodologies into their scientific pursuits.

Each interview is complemented with a technical box that describes an advanced deep learning methodology relevant for the research work of the interviewee.

7.1 Mattia Prosperi on Interventional Artificial Intelligence

Hi Mattia, how did you come into contact with machine-learning and artificial intelligence?

My background is in Computer Engineering (Fig. 7.1). It all started with my master degree on self-locating robots—at the time, they were very simple things, machines with four sensors that kept banging against the walls. But it gave me the chance to use some machine-learning algorithms for localization; we tried to go beyond simple Kalman filters. We started using biology-inspired algorithms.

One day, a professor from the Catholic University of the Sacred Heart in Rome came to our lab and said, "I'm studying drug resistance in HIV viruses"—at the time there were still very few drugs—"and we want to create an algorithm that uses both statistics and machine-learning and that is able to optimize therapies". And so, my professor asked me if I wanted to do a thesis on this topic. I was already very interested in biology and medicine, so I accepted and began to develop these

F. Biscarini, N. Nazzicari, *Deep Learning for Life Sciences*, Decoding Evolution 1,
https://doi.org/10.1007/978-3-031-96852-5_7

Fig. 7.1 Mattia Prosperi is professor at Florida University and associate dean for Artificial Intelligence and Innovation. His research is focused on biomedical modelling and data science. He leads projects studying HIV transmission, HIV treatment resistance, and HIV health/life outcomes, and other projects on antimicrobial resistance, as well as disease prediction modelling

decision-making algorithms, using fuzzy logic, that tried to interpret virus mutations in relation to drug resistance.

This project then led me to a doctorate programme on the study of genotypic resistance and the prediction of in vivo therapy using machine-learning, Markov chains, a set of computer engineering concepts repurposed for medicine.

During my doctorate, I went to the Max Planck Institute (in Munich, Germany), home of one of the strongest bioinformatics schools in Europe. Afterwards, I did biomedical research in Barcelona and at University College London, where I studied statistical learning and epidemiology rather than machine-learning.

I then spent a couple of years back in Italy, first at the Spallanzani Institute, then at the Catholic University, always working in the field of bioinformatics. Meanwhile, the first Next Generation sequencers had come out, so I threw myself into bioinformatics algorithms. One of my most cited works is one of the first assemblers of viral quasi-species—almost no longer used anymore because sequencers now make reads almost as long as a complete viral DNA—but at the time it was a significant step toward solving a relevant problem.

Summing up, my path was all about bioinformatics during my postdoc years, when I moved to the United States. I then took an Assistant Professor position in the United Kingdom (University of Manchester), shifting towards biomedical informatics in general, and finally towards big data. The path was therefore from bioinformatics to data science, health data science and, finally, artificial intelligence in medicine. Which, at this point, is not exactly brand new: we have a journal that has been around for over 10 years.

I finally joined the University of Florida as an associate professor, joining the Department of Epidemiology and incorporating the rigorousness of that field into biomedical informatics, i.e. epi-informatics. There was a bit of difficulty at the beginning because the US is a very difficult country for tenure tracks, but then when the first grants came in, when I began to create a significant research portfolio, I had the chance to do what I liked.

And here we come to the central point: Artificial Intelligence for Health which, to me, is more than just predictive artificial intelligence, as it is used in most instances. I see a central role for **interventional AI**.

Please tell us more. What do you mean by interventional artificial intelligence?
Let's take a step back. For me, when we talk about artificial intelligence we incorporate the themes of self-consciousness and **general artificial intelligence**. However, nowadays, the term is usually used as a synonym for non-parametric application models, that is, machine-learning on a set of data. It is essentially learning from datasets.

Using non-parametric techniques—everything from decision trees to support vector machines, as well as neural networks—learning can occur and a model can be obtained that can be used for predictions. And here we come to the difference between predictive and interventional AI systems.

Predictive AI makes no distinction among data that may be biased, that may have been generated by some underlying mechanism, or that which includes unobservable elements. They "couldn't care less". In epidemiology we talk about confounders and selection bias or biases of various types because, when making predictions, the important thing is to say exactly what will happen in the future without changing anything in the present. Therefore, if nothing changes, we don't care if there is a bias in the data. We are not going to change the world.

An exemplary case of this is ethnicity. Take a snapshot of the world at this precise moment and you will discover, for example, that in the United States there is a huge disparity between ethnic groups regarding cholesterol levels and diabetes. Therefore, ethnicity will be a good predictor of the outcome of a disease. Biological sex, or gender, is another element that greatly influences predictions.

And here we see the difference with interventional artificial intelligence. If we change specific variables, we no longer reflect the picture of the world at the moment in which the data were generated, and therefore we no longer make predictions, we make interventions.

Obviously, there are variables that cannot be changed. However, even when we talk about changeable therapies or behavioral interventions—e.g. following a certain type of diet—it is not certain that activating these variables in the machine-learning model as learned from the data will then result in the predictions we expect. We are investigating alternative worlds.

For example, people who smoke often have yellow teeth or carry matches with them. A machine-learning system cannot distinguish between correlated variables, and therefore it could say that whitening one's teeth reduces the health risks associated with smoking.

Let's take a different example: changing your diet to improve your blood pressure. A diet high in fiber and without ultra-processed foods may not be viable in certain demographic groups in the United States. There are areas called "food deserts"—mainly rural areas—where it is very difficult to find a supermarket. There, the population has access to food only in convenience stores, where most food is ultra-processed.

So, that variable—changing a patient's diet—is absolutely not actionable, and it is likely that a machine-learning system that learns that type of data is not even aware of containing that bias.

You are saying that there are two elements: the classical, perhaps naive neural network, which simply records correlations without understanding the causal relationships—as in the example of smoking and yellow teeth. The other is the more or less extensive knowledge of the out-of-set examples, i.e. the reality outside the training dataset.

There is also a third thing, namely the availability of measures that can help us find the causal relationships we are looking for. My research works on all of these three fronts.

For the first one I use the classic causal inference of epidemiology, but there will be many other approaches that come from Computer Science. And what I usually do is just impose causal constraints on deep learning models.

The second one is to understand how to acquire data sources that can help us measure and collect variables that can then be actioned—in fact, often the problem is that there are variables that are not actionable but which still influence the causal path and can be measured. And even the boundary between actionable and non-actionable is sometimes blurred. There is always the possibility of having, for example, programs at the local community level, or even state programs, that increase or improve education, or welfare. And those initiatives have a causal function that can then help compensate for demographic, ethnic or racial disparity.

The third is to test these new algorithms on out-of-distribution cases, i.e. populations different from the training ones, and see how the algorithm generalizes.

But in practical terms, what do you work on?

As far as applications go, we work on infectious diseases, in particular HIV and bacterial infections, and mainly on epidemiology studies. This is the research I am directly involved with. Then, as associate Dean for the College of Public Health Professions, I work on three fronts.

The first is public health research using AI. The mission is precisely to solve the problems that we were not able to solve before, to do research that was not possible before.

The second is to train the new generation of scientists in a way that goes beyond traditional training in public health and epidemiology. Here, there is a bit of everything, from physical therapy to speech therapy, and we cover all healthcare professions. We are lucky because we also have the biostatistics department, very strong on a quantitative level, which provides data and expertise from a methodological point of view.

We do have some challenges, though. For instance, developing, maintaining, and making big biomedical data accessible for research and public health implementation. Real-world biomedical databases are very complex, and they require well-governed, specialized informatics capabilities.

Do you mean server maintenance or strictly programming?

I would say data infrastructure and middleware programming. All medical databases are maintained by the College of Medicine. The data infrastructure of the University of Florida is exceptional. We have a consortium called "One Florida" which covers over 60% of all hospitals in Florida and also, more or less, the patients. We can link to ecological databases, so there's everything from particulate measurements to what the average income is in a given neighborhood.

When we have scientists who need to develop whole platforms, apps or intervention programs, we need to find the right collaborators across our colleges or external partners to make sure that they are handled the right way. Part of my job is in fact building these bridges and making them self-sustained.

On the other hand, if we need to write the equations for a deep learning model that does debiasing, we can do it and provide such expertise to others.

What about training?

I have developed two certificates, undergraduate and graduate, in Public Health AI. They are not exactly short degrees, but something similar, on the topic of public health and AI. It is not our objective to create programmers or computer scientists. We rather aim to train the public health scientist or epidemiologist who knows how to access medical databases with standardized ontologies, who knows how to train advanced models, and who is strong in biostatistics and causal inference and well versed in AI tools useful for their job. This is the new generation we are trying to create.

Wasn't there a third front?

True! The third front with respect to innovation is to increase relationships, especially in terms of research, but also in terms of the influx of our graduates into the industrial sector. Historically, our university had primarily focused on federal or state funds while, in regard to my personal experience, I have also had great satisfaction in working with the industry. I have worked with several companies, both small and large, that are deliverable-focused, and there are far fewer frills. It is also more satisfying to compete for funds from a certain point of view because, if you have the right expertise, they will come looking for you. Even seeing our students go to work directly for a company after graduation or after their doctorate—sometimes with salaries even higher than ours—well, it means that we have done a good job in training them.

We have a very strong partnership with Nvidia; the University of Florida has among the largest GPU clusters in North America or in all of America.

Here at the College of Public Health we also work with pharmaceutical companies. I have personally even collaborated with chemical companies because they were working on herbicides, pesticides, and were trying to evaluate the perception of health risks in the population, as well as to understand their reputation among the population. It was a very interesting study, using Natural Language Processing in social media.

Let's move on. Has the advent of Deep learning, as a technology, represented a discontinuity in the researcher's profession? Or was it just a new step in an already well-defined trajectory?

There are subfields in health in which deep learning has been truly transformative. We've seen the examples of protein structure prediction, for which it basically destroyed more than a decade of research.

Connected to this is drug design. One of the first antibiotics created via deep learning was recently synthesized and will be able to be tested in vitro and in vivo. The problem in that area was that, in the beginning, the models generated non-synthesizable structures; they were just too complicated. But now they are getting better, simpler solutions.

Other examples come from radiology, image interpretation and the like. These are all subfields and specific applications for which deep learning has been truly revolutionary. It is also starting to find applications with the analysis of electronic records and real world data, even if there was a significant disappointment there in previous years. Many articles have been published along the lines of "Deep learning of electronic records can predict everything", and then the models did not generalize: you obtained predictions that made no sense from a clinical point of view. Things are still evolving in that field but. at this moment. I am convinced that things will change a lot with **Causal Deep Learning**. However, this scares me a little because when using Causal Deep Learning you still have to make very strong assumptions about causal relationships. And if those are wrong, the models continue to be wrong.

Are you referring to Judea Pearl's work on the introduction of cause-effect links in neural networks, with Bayesian networks and subsequent developments?

I work on introducing causality; we can call them Bayesian networks, he calls them DAG—Direct Acyclic Graphs. I use this approach to introduce prior knowledge. Those structures need to be given to the model to be used as a deep learning constraint.

There are other approaches, however. For example, Peter Spirtes at Carnegie Mellon University worked on learning causal structures. Unfortunately, the problem there is that it has been demonstrated that, given a dataset, there is more than one possible causal network, all of which have all the same maximum probability relating to that dataset. And therefore it is practically almost impossible to obtain the ground truth of causal relations with certainty.

But let's not despair. No one gives us humans causal networks, but we discover causal relationships nonetheless, usually through trial and error. Then, when we want to be a little more formal, we do random screening or randomized clinical trials. But in any case we gain intuitions by observing the data, that is, by observing reality, which can and will be biased. And that's why we make mistakes. I don't see anything different in learning multiple causal structures, and therefore having multiple hypotheses, which then need to be confirmed with other types of data or evidence.

On these topics, deep learning is promising, but we are not at the revolutionary levels we have seen in radiology or drug design.

What else is there? I guess I should mention language models.

Well, we are in the era of language models.

I use them to help write letters of recommendation, fix the language and grammar in write-ups, other things. I still don't use them to write research proposals and similar things because—besides what agencies allow—I think, let's say I have the arrogance to think that I still write better than them, especially because my specific knowledge of the topics is probably better. But for generalist tasks they are certainly a revolution. We have moved from paper encyclopedias to Wikipedia and now there is chatGPT, which has encyclopedic knowledge and explains the facts in simple words. And this is a paradigm shift towards knowledge.

In health research, we have language models that are developed on clinical notes. At the University of Florida we have Gatortron [125], which is the most powerful language model with the most parameters in the world that has been trained on medical data. It's very interesting, because it helps to make medical queries about patients, it helps with core discovery, with analyzing the data. Clinical notes contain a lot of qualitative, or semi-quantitative, information that large language models can incorporate. For example, they are very useful for selecting patients (i.e. checking eligibility criteria) for specific therapies, or when designing a trial or creating an observational study, for when you want to extract measures that are not standardized or incorporated into ontologies. So, this could also greatly facilitate the collection of medical data, because clinical notes are more natural for a doctor than having to spend 20/30 minutes after each visit encoding the diagnoses.

The projects and applications you are working on give us an idea of where the sector is going now. But let's take it a couple of steps further. Every technology makes its own U-shaped curve (parabola) and, after a while, it exhausts its innovative drive. In your opinion, where is deep learning in its trajectory today? Are we halfway there, are we at the beginning, or is there still much to discover? And, once the innovation provided by deep learning has been absorbed and stabilized, how will things have changed?

In recent months, a very heated debate has been taking place among artificial intelligence experts, people like Yann LeCun, Geoffrey Hinton, Yoshua Bengio. Listening to these voices, it seems that everyone has their own approach and vision, more or less catastrophic. My answer is, honestly, I don't know. Let's put aside the big issue, whether or not artificial intelligence will cause the famous doomsday scenario. Let's talk about the curve. It is still difficult to understand where we are. I work on it, in practice, and I can't tell you where we are on the parabola. It depends on the speed with which we move along the curve, how easy or difficult it will be to apply the new solutions, things that no one knows.

This also leads me to reflect on the boundaries that are being set due to the ethical implications. I work a lot on the fact that AI can generate harm, because by working on interventional actions you also need to try and understand if your predictions give you results that are realistic. Wrong predictions regarding the consequences of active interventions can have terribly harmful consequences.

So implications are ethics, since we are talking about policies that affect public health. But the same considerations can be applied to the constraints that are forced on the models. You can easily see this with ChatGPT, which now refuses to answer dangerous or offensive questions. The problem is that it's not possible—let's say it's very hard—to produce policies that are appropriate for all nations, for all situations, and so forth. So either using AI without boundaries or limiting it through constraints will have consequences which are unpredictable.

And that's where I stop. I'm not a professor of ethics, I'm not geared for this kind of discussions. I see the difficulties and it's clear that I'm not going to solve them all! What I could offer on this topic stops being my educated opinion and becomes just my personal opinion, and that doesn't count. In fact, it should not count.

Fine, the world of AI is full of gurus and experts, it's OK to take a step back. But if that is the case we have another question for you. Let's close our eyes and forget about all the technical limitations. What is your dream application for AI?

I give you a very practical answer. First, you need to keep in mind that efficacy and effectiveness are different concepts, and it's important to be able to distinguish them and shed light on them. With that in mind, my answer is drug repurposing. That's a very big topic. We have many drugs developed for specific conditions that may have a lot of potential benefit for other diseases. So, if a genius student comes to me saying "look at this, with this network"—probably with this causal network—"we learn everything on a drug, its efficacy, its effectiveness, the side effects and the effects on illnesses different from the one for which it was developed", well, that day I'll be a very happy scientist.

That is my dream application. It's very practical, and while I'm not saying that's already doable, it's not impossible.

The Problem of Out-of-Distribution Samples

The expression "out-of-distribution (OOD) samples" refers to data points that differ significantly from the data to which the model has been exposed during training. It's a problem that happens during the inference phase, i.e. after the training is completed and the model is used to perform actual predictions. Feeding the model with data that deviates substantially from what it has been exposed to during training may lead to unreliable results.

The main source for OOD samples is distribution shift. In the simplest case, the training data doesn't contain (and maybe cannot contain) the full spectrum of cases encountered in real world applications. As such, an OOD sample is bound to pop up due to unforeseen circumstances.

(continued)

A more subtle case happens when the real world conditions change. This may be due to underlying trends (e.g. climate change) that slowly alter the scenario and make it differ more and more from what was described by the original training data. Another source of change may be the model itself. If the network is used to implement some changes (e.g. interventions, policies, medical treatments) these can in turn influence the reality that the model is required to predict.

A slightly different source of OOD samples is adversarial attacks, i.e. usually small, carefully crafted perturbations deliberately added to input data with the goal of causing a deep neural network to misclassify or make incorrect predictions. These perturbations may be imperceptible to the human eye but can have a significant impact on the model's output. One possible approach to **adversarial learning** for classification networks (training a model robust to adversarial attacks) is the Fast Gradient Sign Method (FGSM) [126], which involves taking the gradient of the loss function with respect to the input and then perturbing the input in the direction that maximizes the loss. The process is then iterated until the network changes its output. If the perturbation is small enough the original sample would still be classified correctly by a human expert while the network output has now been corrupted.

Mitigating the problems linked to OOD samples is difficult, especially in general terms. The proposed solutions are discussed in the wider subfield of "uncertainty estimation" [127]. An easily applicable strategy is to implement "Test-time augmentation" [128], in which data augmentation is applied to the test set and the predictions are analyzed. Data augmentation is used here to simulate slightly different but still realistic conditions. The core idea is that the augmented test set better represents the wide variety of conditions that can be encountered in real world applications. This technique has been shown to improve model calibration for both in-distribution and out-of-distribution data. Unfortunately, data augmentation is very application-dependent, with no clear boundary between realistic and unrealistic data perturbation.

7.2 Roland Roller on Building Tools for Medical Decision Support

Hi Roland. Can you tell us how you came into contact with machine-learning?

My background is in computer science and computational linguistics: it was not a straight path to ML (Fig. 7.2). At some point in my lifetime, I decided to embark on a PhD at the University of Sheffield in the UK. I was interested in language and language interaction, but I kind of ended up getting this scholarship for information extraction from biomedical literature, and I happened to like working in the biomedical field. The project was on English text data, freely available

Fig. 7.2 Roland Roller is Senior scientist at the Speech and Language Technology group at the German Research Center for Artificial Intelligence (DFKI). His research focuses on natural language processing and machine-learning with a high interest in the biomedical and clinical domain

biomedical records, and we tried to extract specific sets of information—the details don't matter at the moment, let's just say that it was text processing on large datasets.

When I finished my PhD, I was looking around for job positions, and what happened next was all the result of coincidence. I went back to Germany and applied for a job, but it turned out to be not the right thing for me. Then somebody pointed me towards a colleague whose team was working on clinical data from a German hospital. They wanted to do information extraction. It was not the same as what I did during my PhD but I found it interesting anyway and that's kind of how I started down my career path.

It was at that moment, I think at the end of 2015, when I started working on a research project together with the Charité hospital in Berlin and other partners. It was very interesting, but also very challenging. Before that project I only worked on English data, which was freely available and very abundant. And now, I was working on clinical data, which is locked in closed repositories of sensitive data. Moreover, it was in German, and at least back in 2015–2016 there were just a few repositories of clinical text data in English and that was it, nothing in German. It was quite frustrating to be honest. <laughs>

The topic should be deep learning, right? But, at the beginning, there were few people in the field working on natural language processing. And regarding German clinical text data, there were no resources. So without any training data resources or tools you have to start from scratch, and that's what I tried to do. We tried to create some data sets, and since the real medical data were not accessible we moved toward what was publicly available. Luckily for us people share information, obviously on medical forums but also on public social media. It was a good opportunity to do something meaningful, and that's how I ended up doing what I do. I'm working on health-related topic, I apply AI technologies, i.e. natural language processing and machine-learning to process text. We do identification of adverse drug reactions from social media with multilingual models, and we focus on clinical decision support in combination with **explainable trustworthy AI**. The goal is not only to create a good AI system: we recently published a paper arguing that performance is not enough, because people need to understand what the system is recommending, and why, and they need to trust it. And this is actually even more crucial.

So, you started mainly on language processing and now you are working at the German Resource Centre for Artificial Intelligence on the topic of decision support, correct?

Yes, clinical decision support is one big aspect of my work, it runs through a lot of different projects in which I'm involved. But standard natural language processing is also another big part.

What does a decision support project look like? Can you give us an example?

Sure. In one of my first projects, I worked with the nephrology department. The most crucial thing regarding kidney disease is transplanted kidneys. There are not that many kidneys available but there's a great number of patients waiting, so you really don't want to mess up. Now, transplants are always a risky operation, and doctors want to make sure the new kidney will last as long as possible in the receiver patient. They came to us and said: «*You know, it would be really good to have a system that estimates the risk of rejection*». They actually came up with three separate classes of risk and asked the distribution of probability of survival in the 90 days after the transplant for each of them. And that's what we did.

As you can see, it usually starts with a request from medical experts, who need to provide the idea. Then, you need to formalize the problem. And then, you need to get access to the data and have a lot of close interactions with the medical experts because there's a lot of hidden knowledge in the data. And then things become complicated. As a data scientist, you wish for a simple scenario: just take the data, put it into the model and that's it. But that's just a dream, it's never like this. Data is obviously noisy, but there's more. If you have longitudinal semi-structured records and the database develops, say, over 20 years, you really need to put a lot of effort into which information should go into the model.

You need to consider that I'm a very pragmatic person. I consider technologies as tools, and you should pick the tool that best fits the task. At the beginning of my career we didn't have a lot computational power, and deep learning simply wasn't available yet. We had expert systems that performed adequately, and actually they are still there. Today state-of-the-art models certainly have their advantages and probably outperform simple models, but in order to have a baseline it always makes sense to start with a simple model. You can put one together very quickly and use it as a reference. Also in my experience working with experts, you are continuously asked to integrate more and more aspects of their knowledge, and everything seems to be super important. So the model becomes very complex very quickly.

Also consider that when I started there was no reference paper for this kind of risk prediction. And even if there had been one it would have been problematic anyway, because the moment you test your model on a different dataset, on a different cohort, there's no real comparison.

So you sit there doing black box engineering, you get some results, and wonder: is this good? You don't know. Can this help anybody? You don't know. So at some point we said, «Ok, I can feature engineer as much as I want, we can test more fancy methods, but we need to understand if what we have is good enough to actually support physicians». So we decided to do a small study with physicians and check how effectively we can actually solve this fictitious task of estimating the risk that something bad happens to the patients in the next 90 days and, on top of that, examine whether a physician is better off with our risk estimation.

Thanks for the clear, if stark, description of the difficulties of this kind of work. I've seen one of your recent project was about surveying Twitter data, right?

Yes, we also work on social media data regarding people publicly sharing medical information. It's a collaboration project among Germany, France and Japan, because we want to work with data from different languages, not only English. We want to develop multilingual models that are able to deal with the different languages and detect unwanted adverse reactions.

In this case, you would think that it is going to be much easier, and, to a certain extent, it is, since people publish the data online freely. But often it is not: you can either have very interesting but still anonymous data from health forums, or personal data protected by legal constraints. You still need to get permissions, and it becomes very complicated very quickly. But we manage, and anyway in this case we can we have more computing power and use state-of-the-art models like the transformer models.

During my PhD, I was still working with Support Vector Machines and similar techniques and then, suddenly, the big hype about deep learning started. I mean, it wasn't totally new but it abruptly became a lot more visible due to various reasons. LSTM started, and suddenly everybody was just working with LSTMs. Then some years ago suddenly transformer models popped out and for the last few years now everybody has been relying on transformer architectures and all these thousands variations that are optimized for specific texts. It comes in waves. Right now, we currently seeing the next big wave coming. I mean, without doubt ChatGPT is on the table, but it certainly has big disadvantages. It is essentially hosted by a company, it's not an open source model. So, if you are talking about clinical data, we don't want to put sensitive data into this kind of model. But other models will follow. There's currently a lot of stuff going on.

Let's focus on deep network models for a moment. What are the major benefits and the major limitations compared to other available methods?

The pros are easy, they often provide better results. You've seen so many papers showing a transformer model outperforming traditional models, that it almost doesn't make sense to use anything else—even though I just said that it's important to do a quick implementation of a baseline model (and I maintain this). Sometimes, though, it really doesn't make sense to use a simple model if you already know that the transformer model will be better.

There is also an advantage in terms of the feature engineering, in the sense that it is often simpler to just put the text data into the model without doing any prior "engineering".

Unfortunately, it's not always that straightforward. Sometimes, you still need to do feature engineering—the good news is that ten years ago it was much worse. One mixed topic is explainability, and here it really depends on the problem.

Then, there's the problem of hardware and resources, so maybe you discover that you cannot just take any state-of-the-art model, because you simply cannot train it.

And also data availability. You need a lot of data to train these big models on your own and, especially if you work with clinical data, it's just not going to happen.

And even if you have the data and the resources, you may end up discovering that it's not that easy—otherwise big players like IBM would have already solved the problem. I believe you still need to understand the data and you need to know the limitations of the problem.

For instance, I was working on the data from an emergency department in regard to automatic triage and triage support. I was testing some models and I wondered why my model made a certain prediction. The label was different from what we expected, so I had a look into it and my model seemed to be right. I ended up having some deep interactions with the physicians and I discovered that it was written in the clinical text that the person in the ambulance blood sugar level that then changed when the patient was admitted to the hospital. In fact, the level at the triage stage was much lower, and thus required a more severe classification. You cannot just point a model toward your data, you still need to understand what's going on and know how to treat the data properly. It's not easy.

Now, deep learning. There's a lot of potential, I believe in this. But I think it's more of a tool to make us work better. Hopefully.

How do you see deep learning being applied in your field in the next five years?

In hospitals there's a lot of potential. Change happens slowly for many reasons, but there's a great deal of potential for saving a lot of work for employees. Helping in analysis of the data, detecting possible risks for patients that may be overseen by physicians, those are really interesting applications. Also, there are plenty of boring activities, like coding the hospital's billings. There are already some kinds of automatic tools that run over the text data and estimate the invoice. In general, I hope to see this kind of applications, making processes in the hospital more automatic.

Let's move further into the future. Where do you think this thing is going?

First of all, a little background. Assistant physicians, young physicians, in my experience, they can better detect the risks for patients when working with a machine-learning model, whereas the senior physicians that I met, they kind of couldn't. Actually, they slightly decreased in performance according to our measurements. This was surprising and interesting. Our interpretation was that the senior physicians have a lot of experience and need to have strong opinions because, normally, the young physicians come to them to get advice. Also, they may be very critical about certain details and, if something is not like what they expect, they just don't accept it.

This is why we currently have a research project about how to create a trustworthy AI. It's not enough to get the young ones, we want to get all physicians on board. And so we need to have, for example, adjustable explanations according to the needs of the doctor. Some may want to have a particularly high precision or high recall, or maybe they just want to have some specific information in the model or want to highlight specific aspects. And this is the direction where this is going. Step by step, it will take some years, and young people will probably accept it more readily than others. But if it is proved by clinical studies that the model is working very well, it will be accepted eventually.

On one side, this is reassuring, we want to have evidence-based medicine. On the other side, clinical studies are slow—depending on the application they go from slow to really, really slow, especially compared to the current rate of evolution of deep learning technologies.

There will be always some lag. But also consider that the current evolution trajectory for AI is not clear. Currently they are asking for more and more computational resources, so the top tier, the state of the art, is always too big, too much for anybody except the biggest players. But maybe at some point we will reach some kind of stable state and then the field of medicine will acquire it and include it. And how far will this plateau be?

Many years ago, people were talking about the next AI winter, when it would come and crush all our hopes. And for sure we are certainly in a very hyped phase. I remember I gave a lecture last winter semester, in December (2022, nda), I believe it was the last lecture of the semester and I was talking about state-of-the-art models; it was about that time that ChatGPT was really starting to take over, and we were talking about it, just briefly, with the students. I thought at the time that maybe it was all just hype. But then it didn't really stop, and when for the Easter break my mom visited me, even her—she is above 80—even her had heard about ChatGPT. I found it crazy. The expectations are high but the potential is still big.

And that's true for machine-learning in general, there is a lot of potential and we will see benefits in different areas.

Still talking about the future, what is your dream application? If you could dream of one tool that is not there now, one dream application for artificial intelligence that is difficult to predict or implement today, what would that be?

If I think about health, I think about simple solutions that don't necessarily need to be deep learning solutions. To give an example, some weeks ago I wanted to make a medical appointment to check my eyes, and it turned out to be an incredible experience. It's 2023, and it's still incredibly difficult to get an appointment, I had to go on the website, I had to confirm that I'm already a patient of the doctor and, had that not been the case, I wouldn't have gotten an appointment at all. So, there's a lot of potential to be solved in terms of digitalization and automation, simply streamlining the act of going to a doctor and getting a timely appointment, that would be a big improvement. Not really deep learning, just digitalization/automation. Maybe big data is a keyword, but having this data connected is crucial, and difficult. We have data out there, maybe recorded and stored by our standard family doctor, but also by all these specialists whom we might visit, and they are locked in different silos that can't be accessed. And then, we have insurances and other sources. If the data were available we could actually train incredibly interesting models that would be a benefit for our society. We could draw conclusions, if we connected our health data. And if we do it for the benefit of the public at large, not just for the sake of big companies, if it could be used by everybody, by scientists for public good, I think this would be a great achievement. So, my dream application is not about deep learning as a solution. It's more about the need to move forward with digitalization and automation. There can be a lot of ecosystems and improvements in general.

We talked a lot about the potential of artificial intelligence in the future, but what are the main obstacles that you see to its development?

Maybe the main problem is the human itself. The expectations.

I mean, if we talk about research, the main obstacle is data access. We normally only have small datasets available, and maybe if we work on clinical data and are lucky, we may amass to the data of one single hospital. But if you had access to many hospitals in a large network, and maybe connect to all the other health status indicators, it would be so exceptionally beneficial. Then again, if I think about the legal aspects it quickly becomes a nightmare. And I don't know if this approach of total data sharing would necessarily be a good solution for society in general. But from a researcher perspective this would be great. I can understand people's concerns, though, especially if you don't know what happens with your data. Nobody wants to have their private data publicly available.

OK, so we are about finished now. I only have a very small question before we stop recording. You talked a lot about the analysis of text data and natural language processing. Do you also have experience with deep learning using other types of data like images or sounds or tabular data?

Mainly, we work on text, it's our main thing, and then structured information or longitudinal-structure data. I haven't worked on other data-types yet but I think this is going to be an important aspect. It's what we call multimodality (i.e. developing neural networks that can process multiple types of data, e.g. text data plus image data).

In the past, many machine-learning models focused on one modality, like only text, only time series, only images, and they considered each modality independently. But the information overlaps and expands naturally to different modalities. You see, I'm always talking from my perspective regarding my use cases, and maybe it's not the same for all other use cases, but let's assume I have patient data and I want to model something, any kind of medical application—let's stay very general, let's say I want a task to help humanity, right? In a hospital I have text data, structured longitudinal data, and image data. And so, I will have at least three different models, maybe more. But it doesn't make sense to train different models, because all the information is interconnected. Multimodal models are bound to be an important topic, at least in medicine.

The Encoder-Decoder Architecture

Encoder-decoder architectures are typically used for tasks involving sequence-to-sequence transformations. The most prominent example is probably machine translation, but many other problems related to natural language (e.g. text summarization, image captioning, and so forth) are solved using this kind of architectures. A key feature is the capability of encoder-decoder networks to support variable length data, both as input and as output.

As the name implies, these architectures are roughly split into two halves. The first one, called an **encoder**, receives as input the original sequence and transforms it into a fixed-length internal representation called a **context vector**. This step has been performed using recurrent neural networks (RNN) such as LSTMs or GRUs (see Sect. 5.5) or, in more recent times, transformer models. The aim of this step is to condense the information into a more manageable form, so that it's ready to be input into the second half of the network.

It's then the turn of the **decoder** to transform the context vector into a new sequence, which is output step-by-step (e.g. each element of the sequence at a time). This second half of the architecture, too, is implemented either via recurrent neural networks or transformers. During training, the decoder is fed the correct output sequence at each step. During inference, the decoder becomes *auto-regressive*, i.e. it consumes the previously generated symbols as additional input when generating the next one.

Transformer models represented a big improvement for the performances of machine translation tasks. They were introduced with the seminal 2017 paper "Attention Is All You Need" [129]. The general framework is that of an encoder-decoder architecture, but transformers are based entirely on self-attention mechanisms, eliminating the need for recurrent or convolutional layers. To be fair, **attention mechanics** were not introduced with transformers, having already been around for a while [130]. Attention is a way to circumvent the bottleneck represented by the context vector, allowing other parts of the sequence to "jump in" and contribute if they are deemed to be relevant to the current prediction step. A key intuition in the design of transformers has been the use of (several) attention mechanisms combined with a relatively simple network, thus avoiding the notoriously onerous computational needs of recurrent networks.

7.3 Alex Hernández-Garcia on Active Learning

Hi Alex. Will you briefly introduce yourself and the Mila lab where you currently work?

My name is Alex Hernández-Garcia (Fig. 7.3). I am a postdoctoral scientist at Mila, the Quebec Artificial Intelligence Institute in Montreal, Quebec, Canada. I

Fig. 7.3 Alex Hernández-Garcia is a postdoctoral researcher working with professor Yoshua Bengio at Mila—the Quebec Artificial Intelligence Institute in Montreal, Canada. His research focuses on projects that seek to have a positive social and environmental impact using machine-learning

work with professor Yoshua Bengio. Currently, I'm working on applications of AI and machine-learning for accelerating scientific discoveries to fight climate change. In particular, that means the discovery of new materials, but also drug discovery, molecular sciences, and a lot more!

OK, so your field of application is mainly life sciences

Life sciences and also a lot of materials, engineering, chemistry, crystals, molecular discovery, etc. Also biological sequence design, etc.

And you did your PhD at the Mila lab?

No, I did my PhD in cognitive sciences in Berlin in Germany. It was a PhD at the intersection of deep learning, machine-learning and computational neuroscience: I studied the fundamental aspects of learning in both brain and machines and the interactions between the two fields, namely how AI can be used to further our understanding and knowledge about how the brain works, in particular the visual side of the brain. But also the other way around, i.e. how neuroscience or the understanding of the brain can improve machine-learning algorithms. Before that I completed Bachelor's and Master's degrees in Madrid, Spain, where I grew up. I moved to Montreal for my postdoc at the end of 2020. So, I'm in my third year now.

Just a question on the Mila lab in general. You mentioned your specific areas of work but your lab works on all fields of applications of neural network models and machine-learning models, and also on the development of new architectures, new methods. So it's a more general type of laboratory, right?

Mila started as a group led by professor Yoshua Bengio and then it expanded into a bigger lab, recruiting more professors and staff. Currently it's a multi-university institute meaning that participants include faculty members from multiple universities, such as the University of Montreal, McGill University and others. Currently, because it's grown a lot, the research at Mila spans a wide range of scientific areas. The main focus is artificial intelligence and research in machine-learning, but there's a lot of adjacent fields that are also part of the topics that are studied at Mila. For example, there are professors doing robotics, computational psychology, computational psychiatry, and recently there have been efforts towards social good, meaning we are working on other sciences which, in my own case, means materials discovery, drugs discovery, and so on.

There is both research at the application level and fundamental research about improving methods for machine-learning, neural networks, causality, different techniques like natural language processing (NLP), computer vision, reinforcement learning. So a lot—maybe most—of the areas that are currently studied in machine-learning are part of the research portfolio at Mila.

Let's dig deeper into your current research

I would say that about five years ago we had the explosion of the adoption of machine-learning for drug discovery. That doesn't mean that there was no research on the topic before nor that we did not use any machine-learning. But around four-five years ago, the whole thing started to really come together. This was due to multiple causes, but it was mostly the maturity of the machine-learning methods. I think that's the main cause, ML methods started to become really solid and were able to be used on a large scale. On top of that, there hasn't been a lot of work in the area of materials discovery, which is an area that could be quite impactful in the fight against the climate crisis. There is a huge need for discovering new materials to improve the efficiency of a lot of industrial processes, like for example electrochemical reactions which are responsible for carbon emissions around the world.

So, I started working in these areas because I was interested in doing some research that could have an impact in tackling the climate crisis, and in particular I've been very interested in accelerating the exploration of new materials, new drugs, new molecules. I specifically work in an area of machine-learning known as **active learning**. It is very related to the field of **Bayesian optimization**,[1] in other words very similar to a case that may be familiar to you. The idea is to have a learner, an algorithm that is going to be trained with data. But there's a fundamental difference compared to more classical machine-learning algorithms, in which the objective is to make a model as accurate as possible. Here, what we want is an algorithm that is able to tell us which objects, which candidates, we should annotate next.

The annotation of examples is very expensive in general. So, imagine if we want to characterize a material or a drug: we need to carry out wet lab experiments, or engage in very computationally demanding computer simulations in large clusters of computers. The idea is that the machine-learning methods can explore the space of drugs, molecules, materials, etc., in a more efficient way than a team of scientists

[1] Bayesian Optimization (BO) is a technique for **hyperparameter optimization**. It improves the search speed using past performances (**Bayesian updating**), while random or grid searches are uniform (or independent) of past evaluations. BO is somewhat like manual search: when you are manually optimizing the hyperparameters, you would first try a set of parameters, then look at the result, change one of the parameters, rerun, and compare the results, so you know if you are going in the right direction. Bayesian Optimization does a similar thing: the performance of your past hyperparameters affects the subsequent decisions.

can do. We can train big models and create algorithms that can explore that space of possibilities.

If we understood correctly, in the space of all possible candidate molecules, the output of your predictive machine is a list of three, five, a hundred possible molecules that are of interest. In this aspect it is different from the typical machine-learning application in which the sample enters the model to receive a score or a classification.

Correct. Active learning is different in the sense that we are adding infrastructure. We are indeed using the more classical model that makes a prediction about a candidate. But here we are adding an outer loop that will use this model to explore the large space of candidates.

Keep in mind that it would not be possible to do a complete exploration of the space because it's huge, so knowing where to test becomes relevant. And that is the key. I'm especially interested in problems in which the search space is really huge. For example the estimated number of small molecules that can be used for drugs is on the order of ten to the sixty (10^{60}).[2] So, even if you have a super-fast machine-learning model, you just cannot explore even a tiny part of that solution space. What you need is smart algorithms that can jump all over the place and do smart exploration of the space of stable materials.

So you're trying to segment the space of possibilities into areas that are potentially more interesting?

Even better than that: we aim to discover and develop algorithms that can actually guide the research. Let's talk about molecules for drug development. I think it's a more intuitive example. Maybe we are looking for a drug to cure a specific virus-disease, like COVID-19, and we are looking for a specific property, like the binding affinity with the virus proteins. And maybe we have some data points that we have characterized in a region of the space. So what we want is an algorithm that can extract high level properties of the data that we have and say, «Hey, in this far away region of the space, there are certain similar properties that might be useful for what we are looking for, that might potentially have high binding affinity or some other desired properties». The loop ends with characterizing the candidates that the algorithm has selected with what is called an "Oracle". It could be a wetlab experiment or a computer simulation, but it's always something slow and expensive, and it provides new ground truth about the selected candidates. We then feed in the new information to retrain the predictor and start the loop again.

It's this iterative process that we keep improving. It's a little bit like what we do in a traditional scientific discovery pipeline, but instead of being driven only by hypotheses from scientists, the hypotheses are now proposed by a learning algorithm, by the network.

---

[2] To put this in perspective, the estimated number of protons in the observable universe—the **Eddington number**—is 10^{80}. A **googol**, which is used to illustrate the difference between a difficult to imagine huge number and infinity, is 10^{100}.

Do you have some type of specialized neural network architectures for this type of task?

No, not really, because the innovation here is not at the architecture level. The architecture is going to be that of the models that characterize the candidates: it is going to be the typical model that we would use for the classical algorithms, and it should be tailored to the task at hand. If we are working with molecules, then we want to use graph neural networks (GNN); if there are images involved, we will be working with convolutional neural networks (CNN), and so on. The innovation here comes from other aspects.

One crucial part is the acquisition function, that is, the criteria by which we want to select the candidate. Typically, this is based on information theory; we want to maximize the mutual information about our estimated posterior value of the black-box function that we are trying to optimize.

But since you are asking for specific architecture, what professor Yoshua Bengio's group has recently proposed is a new learning algorithm: it's called **Generative flow networks, GFlowNets** for short. These algorithms are designed to explore the space of candidates efficiently. These algorithms can be seen as samplers, with the property that they sample diverse candidates. This is something that previous algorithms couldn't really do. As an example of a sampler that performs optimization, we have classical Markov Chain Monte Carlo methods that are good at doing the sampling, but they really struggle at mode mixing. So, they can find all the modes of an underlying black-box function, but they can only do it at exponential time. And this is really not a good property.

Could you mention an example of such projects, in which you used this kind of technique to explore a very huge feature space?

We have applied this kind of method quite extensively on biological sequence design. For example, the design of DNA aptamers or antimicrobial peptides, and thus proteins. Here the problem is a little simpler from the modeling perspective than, for example, materials discovery, in which we want to characterize crystals that are more complex objects. Proteins are a sequence of amino acids that we can represent using a string of letters, and there's already been some work on proposed neural networks that can, for example, characterize the toxicity of a sequences or its binding affinity with specific targets.

In this field, we used active learning methods and found that GFlowNets are particularly effective because of one shortcoming—and this is quite crucial—of existing methods. For this kind of problems, the workhorse algorithm has been **reinforcement learning**, which can be seen as a family of optimization algorithms as well. And they are really good at finding the optimum of a black-box function. Their limitation is that they get stuck on that optimum point and if that's all you need then you're good. The problem with this approach is that, typically, what you are optimizing, for example the target binding affinity, is not going to be enough for your downstream application. Because you're going to test your candidate with, let's say, non-human animals and then humans, often you are going to find that there are actually unexpected downsides unaccounted for by the model: maybe the molecule is toxic or it has other unwanted effects. And what you really need in order to overcome this limitation is to propose a diverse set of candidates, right?

Reinforcement learning methods and MCMC and other typical machine-learning algorithms have the problem of not being diverse enough in their proposed solutions, and this is where the difference of deep learning really shines. In my case, I've been using this difference inside active learning loops and particularly for something called **multi-fidelity active learning** [131]. This covers the case in which we have access to multiple oracles: we can characterize a protein in a wetlab experiment, but we also have access to computer simulations, for example based on density functional theory, or other methods and approaches that are less expensive, less costly, but also have less fidelity.[3] There has been a surprising lack of research on how to leverage the availability of multiple sources of information that have different levels of fidelity but also different levels of cost. By using this difference, we have recently shown that we can increase the cost effectiveness of exploring the design space by orders of magnitude. Intuitively, we make a lot of calls to the cheap oracle to quickly explore the design space, while limiting the computational budget, reserving the use of the expensive oracle solely for refining the exploration.

Let's now move on to the more general picture. What do you think are the main advantages and limitations of using neural network models compared to the classical machine-learning approaches?

I think that the main advantage of neural networks with respect to classical algorithms is their ability to learn from the data without the need to craft descriptors. The feature engineering part that we have always done—including myself when I started research. Feature engineering is effective if you really know what you want to do, but it has a hard time leveraging large data sets, while large neural networks really shine at this.

We can have a big dataset and, over the last decade or so, we have developed algorithms that have very effective inductive biases for different types of data. For example, convolutional neural networks are great for dealing with images, while transformers or recurrent neural networks are great for dealing with sequences or with text, for example. This is something that cannot be parsed, not even close, by classical algorithms. Neural networks can very easily learn from large datasets and generalize pretty well, and now we have mastered how to train them.

The disadvantage is that we really need to have information about the nature of the data, and it is a well-known limitation that, whenever we don't have that, neural networks struggle. An easy example is the case of tabular data. It is well known that classical random forest learning algorithms are a lot better than neural networks at learning from tabular data. For some people this sounds surprising, because neural networks are supposed to be "the best" out there, the state-of-the-

[3] The careful reader might have spotted that this bears some resemblance—conceptually—to **weak/slow learning**, i.e. the idea of combining multiple weak learners into a more powerful, combined strong learner, like what is done in **boosting** (Robert E. Schapire, 1990: https://link.springer.com/article/10.1007/BF00116037).

art. The underlying reason is that there are no strong inductive biases, at least none that we know about yet. There is nothing that neural networks can really exploit from tabular data, because it's very disorganized (i.e. random order of columns), it is a type of heterogeneous data in general (see Example 5.1 for a discussion on structured and unstructured data). Another limitation of neural networks is that, in order to make them shine, we need a lot of data. Actually, this is getting better and better. We are getting better at making them work with low data regimes, and one way we can actually do this is by using the methods of active learning that I was talking about. This is because active learning can be used with a small dataset and even a simple algorithm, and as we collect more and more data, we can train larger networks if needed.

Let's build on your example of small problems with tabular data, which are solvable with traditional methods like random forest and with which neural networks struggle. Do you think that this is an intrinsic limitation of neural networks? Or do you think that, maybe someday, there will be a new paper with a new algorithm, a new architecture demonstrating the equivalence of the solution, both in terms of goodness of fit and computation resources?

I don't think it's a theoretical limitation, I wouldn't be surprised if that paper comes out at some point. However, what I would say is that we know that random forests, for example, are great at this specific task, thus we should really talk about what the limitations of random forests are, maybe like data size, computation speed and so on. But, otherwise, for the cases in which random forests or other algorithms work well, there's not really a big need for neural networks.

For most of the use cases in which decision trees and random forests work well, even if that paper comes out and we find a way to do it with neural networks, well, cool, but I don't think it would be a breakthrough, a game changer. You know, another limitation of neural networks is the interpretability of results. So, actually, if we can solve a problem with random forest, that's great, because it's a lot more interpretable, a lot simpler. In fact, I wish we could use random forest for large problems of image recognition or molecular design, but unfortunately they are not suited for these kind of problems.

Let's broaden our horizons. Where do you see the field going and what do you hope to see?

Well, I might be biased by what I'm currently working on, but I'll try to speak in general terms of the field of artificial intelligence for scientific applications, which is a broader area than what I just work on. There are applications of machine-learning for chemistry, for climate science, for material science, for biology. I have colleagues working on biodiversity monitoring. There are lots of fields in the classical sciences that only recently started to use artificial intelligence, and I don't think it's a very bold prediction to say that this is going to be quite a revolution in science. I think we're going to see an acceleration of the things we can do thanks to the use of artificial intelligence, this is what I would predict. I also have been in contact with colleagues from other fields. Yesterday I had a discussion with a post-doctoral researcher who works in astronomy, specifically, with images from super high-dimensional large spaces. They want to explore, or maybe it's better to say to discover, new theories that explain the universe, the fundamental laws of matter. These kinds of things sound a little bit like science fiction right now; but I think it's really going to be possible during the next decade or so, thanks to the use of artificial intelligence, and precisely the kind of algorithms like active learning that I was describing. Large scale exploration is now achievable. This is the kind of things that scientists need. More than that, we have models that can propose theories. Feed them the data we have collected about the world, be it biological or astronomical, it doesn't matter. The machine-learning models are going to be able to propose theories. And a theory can take many forms: we are used to thinking about it as an abstraction, but it could be a random forest for example, a decision tree could be a theory as much as an equation can be a theory. We are going to have machine-learning models that can propose theories that explain the world with high probability, and we are going to have many such theories, and then the scientists are going to say, «We can test them, we can evaluate them». This is only just starting right now, but I think this is where we are going.

What about those problems that are notoriously difficult to predict precisely, like earthquakes or disease outbreaks? Do you think that deep learning will also be disruptive there?

I have no idea about the science of earthquakes but, just intuitively, I think it will be possible. The predictive power of neural networks has shown remarkable success in multiple fields. So, what we need is data, and then we need smart and efficient ways of using that data. Right now, even if it seems hard to believe—from the outside there's a bit of an aura around AI—we are lacking in methods or techniques that can handle large amounts of data in an efficient way. The good news is that there are a lot of people working on exactly this right now. So, it is really going to change, and this is why I think that, during the next decade or so, we are going to see quite a bit of a revolution in many sciences.

The field of machine-learning during the last decade, more or less, has focused on well defined toyscenarios in which we have a dataset and we train a model multiple times on the same dataset, by refining our model iteratively: by now, we know pretty well how to do this.

Then it came the time to put the models out in the real world and, even if I don't like the expression, use them for real, relevant applications. We thus discovered that the things that we thought worked so well actually did not in fact work so well in many applications. This has sparked a lot of interest and research work in developing these methods further, as there's no theoretical limitation in regards to making them work. It just needs more research, but this is already happening. Just in the last two or three years, I've seen a lot of progress on this.

Since it's going to be super pervasive, do you think that either some applications of machine-learning will be commodified and very well standardized, or that every group will need at least an informatician or a statistician who can use these applications? Obviously you are actually doing research on the models, but people just are going to use the model most of the time.

I think this is already happening: to a large extent the commodification of AI applications and the standardization of techniques and tools for using AI are here. This is already happening, and it has some implications, ethical implications, and potential risks as well. The misuse of the techniques is already happening and, as things progress, it can only get worse. A few years ago you really needed to be a specialized scientist and programmer, a good programmer, to be able to use these methods. However, honestly, right now there are these large language models, big neural networks, that can be used with a few lines of code, you don't need to be an expert. That's already happening and it's gonna happen just more.

You just mentioned the risks associated with artificial intelligence. Where do you position yourself in the ongoing debate between pessimistic scientists like Geoffrey Hinton, who warns against the risks linked to artificial intelligence and neural network models, and more optimistic figures like Yann LeCun, who says that the risks are overstated?

It's a good question. So, my personal position on this is not in line with either Jeff Hinton or Yann LeCun. I mean, with all due respect, I would say that the Yann LeCun's position that says, «Oh no, we are all good. There's no problem», is a bit naive. I would say it's even a bit careless to not consider that there may be actual risk involved, and we need to think about that. At a minimum, we need to think about it. I think that would be a more scientific approach, rather than saying no, don't worry, this is all good. Of course, I'm simplifying, but I hope you get the point.

On the other hand, the concerns of people like Geoffrey Hinton are a bit beside the point, in my opinion, and I'm not the only one saying it. To me, they both are focusing too much on these so-called existential risks, human extinction, long-term risk, this scale of problems. I don't think that that's unimportant at all, it's fine to have thoughts about these topics, but to me the real concern should be more for the short-term risks. Like, for example, how AI can impact underrepresented people, how it can impact minorities, how it can impact our societies, democracies, and what the possible misuses of AI are. This is happening right now, and we should probably be more concerned about how big players in our societies—like a big corporation with a lot of power—how they can misuse this technology to engineer whole sectors of the society. This is something that can already be done and I think we need to think more on that problem, which is more concrete, rather than discussions about long-term problems.

Obviously, existential threats have their own fascination and people like a good doomsday scenario, and so these discussions are going to attract a lot of interest and hype.

But there are concrete risks and threats that are actually already happening. There are books about this, not in regard to the future, things that have already happened and we really need to care about that. That's my personal opinion.

Active Learning

Active learning is a paradigm used to improve model performances in cases in which either data is scarce and acquiring new data points is expensive or the feature space is so big that no sampling could represent it fully. When we say "querying phase", we refer to the fact that the neural network can interactively query the user (e.g. the developer) to acquire information on new data points (labels, outcomes). The process of identifying the most valuable examples to annotate (e.g. label) next is usually called a "sampling strategy" or "query strategy". More details can be found here https://towardsdatascience.com/active-learning-overview-strategies-and-uncertainty-measures-521565e0b0b.

Active learning is an iterative process. First, a traditional, inner model is trained on the available data. Then, during the **querying** phase, a second, outer model selects a subset of unlabeled data points for which the inner model is uncertain or which is likely to benefit the most from additional information. This uncertainty can be measured using various criteria, such as entropy, margin sampling, or uncertainty estimates [132].

Once the outer model has provided a list of desired new data points, an **oracle** is used to label (some of) them. Labelling is usually a slow and expensive process that requires human intervention to perform actual experiments, either in vivo or in silico.

The inner model is then trained again on the expanded dataset, and the whole process iterates until either the performances are satisfactory or the budget for new data acquisition is consumed.

This process stands in contrast to traditional passive learning, in which the model is trained on a fixed dataset without any feedback loop for selecting additional data. In principle, any model—not only deep neural networks—could be improved by applying the active learning paradigm.

The key step of any active learning algorithm is the querying phase (identification of a new example to acquire or annotate), and its implementation needs to reflect the actual requirements of the problem at hand. There are several strategies available:

- **Uncertainty Sampling**, such as *Least Confident* (select instances for which the model is least confident in its predictions, e.g. using softmax) or *Margin Sampling* (choose instances with small differences between the top two predicted class probabilities, since a small margin indicates that the model is uncertain about the correct class) [133].
- **Diversity Sampling** selects instances that are representative of diverse regions in the input space to ensure that the model encounters a variety of examples during training [134].

(continued)

- **Query by Committee** maintains an ensemble of models and selects instances where the models disagree the most. This approach assumes that instances where the models have high disagreement are more informative and uncertain [135].

The choice of querying strategy depends on the specific characteristics of the data and the learning task. Different strategies may be more effective in different scenarios. In practice, a combination of these strategies or adaptive methods that dynamically adjust the querying strategy during the active learning process may be employed to achieve better results.

7.4 Elena Casiraghi on Knowledge Graphs

Good morning, Elena. Please tell us something about your story. How did you start using neural networks?

I started from medical images about twenty years ago (Fig. 7.4). At the time the first models were starting to be developed, the machine was asked to help diagnosticians and radiologists in highlighting particular diseases from the images. I started from the applications that we all knew well at this point, which are lung X-rays, postero-anterior ones, the objective was to identify lung tumors in the early stages. At the time, they were one of the major causes of death from cancer. Many psychological studies showed that, even though tumors, which then became metastases, were visible on previously acquired X-rays, radiologists, mainly due to humans being prone to errors after many hours of work, were sometimes not able

Fig. 7.4 Elena Casiraghi is associate professor and lead of AnacletoLab, Department of Computer Science, Università degli Studi di Milano, Milan, Italy. She is an affiliate researcher at Lawrence Berkeley National Lab and visiting professor/researcher at the Computer Science Department of the University of Helsinki, Finland (in March–June 2023 and March–June 2024). Her research focuses on automatic machine-learning systems for medical and biomedical (image, multiomics, exposome) data processing and pattern recognition, with a current focus on multimodal patient data integration and analysis, biomedical knowledge graphs analysis, and LLMs for protein design

to identify the tumor lesions. Therefore, the objective was clear from the beginning: we didn't want to replace doctors, we wanted to provide help. This research was supported by the fact that other studies had shown that two doctors working together, two diagnosticians, radiologists or similar specialists, made fewer errors. However, due to a shortage of experienced radiologists, double investigation of the same X-ray images was impractical.

Obviously, at the beginning of these studies, in Italy—we are talking between 2000 and 2007—things were pretty different. You didn't have all the open data that are currently available. So, you had to go directly to the hospitals and propose your idea and ask for the data, and the standard reaction was having doctors saying yes, yes, whatever, and not even really considering it. Then I met a great physician (Dr. Vanzulli, in those days head of the radiology unit), at Niguarda in Milan, who immediately trusted me and gave me access to image data, so we could get started with our research.

The incredible thing is that even tasks that are trivial for humans—for example, the identification of the lung edges—are very challenging for a computer. Or, at least, they were at the time. In the field of image processing that meant looking for the best combination of filters. We first wanted to look for the tumor area, and within that area we wanted to highlight zones contrasting with the background. And this is already challenging, even for humans. We basically went through the list of the most common filters from the literature and tried all of them with just a little help from theory. The number of possible steps and combinations became very big very quickly. Also, consider that there's no standard case, so that we had to adjust the procedure for at least the most common situations, small lungs, large lungs, rotated lungs (from the X-ray images).

It was more or less a heuristic approach, right?

Yes, heuristics. You can call them rule-based methods or, as the reviewers have done many times, ad hoc methods. This highlights a possible limit of this approach, that is, they work, even quite well sometimes, on a specific database. But when you apply them to other sets of data, these solutions don't work anymore. In hindsight, this may be true, in the sense that it is true that we tried to parameterize everything and extract the maximum information starting from a situation of data scarcity.

At the time, the first machine-learning models were starting to emerge—support vector machines were very popular. It was a change of paradigm: instead of using a lot of energy in preprocessing, you had these methods that learned like children do. You just had to expose them to examples, over and over again. And we could build on these systems right away. We already had tools that worked pretty well in segmenting the areas and extracting key features, texture, patterns in the tissues. You see, these are still human-understandable features, an inheritance from our previous efforts. We had tons of these data, so the very first application of machine-learning for us was feature selection: we fed all the features to these new models and they learned both to extract the most interesting ones and to use them for predictions.

This is what is now called the field of **radiomics**, a name born about ten years later, in 2015–2017. The general pipeline is: take an X-ray image, have an expert radiologist segment it by hand and, from that segmented part, extract a set of features that describes everything to a machine-learning method, which then does its job, i.e. predictions, classification, the usual things. There is a tool (3DSLICER) that allows the expert to segment a volume: this is recognized as the most difficult part of the job, prone to error and to inter/intra-operator variability, especially in ultrasound scans and similar 3d-image data.

The next step would be neural networks, right?

Neural networks, yes, and any other powerful models that leverage neural networks and that we have today as the state of the art. At the time I extracted the features on the basis of filters known from the literature. The interesting aspect is that neural networks, especially the convolutional neural networks invented by Yann LeCun [136], follow the human perceptual model. In fact, convolutions actually imitate the convergence of the human neural system. If you look at the eye you will see that it is made up of many sensors that collect information called cones and rods. All signals then converge in a specific structure called fovea where multiple cones and multiple rods send their output to a single neuron in the next layer. In turn, each of these same neurons processes the received signal and sends its output to a neuron in the second layer, which in turn receives many of these signals. This is why it is called the convergence of neurons, and it explains many perceptive tricks and quirks, and also those perceptive hallucinations that you sometimes encounter. The nice thing about these operation is that it has a mathematical counterpart—convolutions.

For me, it was fantastic. I was there trying out all the filters and combining them and it was very tiring, and then these new neural networks showed up and did the whole job for me!

The arrival of neural networks implied that you no longer needed this first, labour-intensive, hardly generalizable preprocessing phase. You can now start directly from the raw data, or at least much closer to the raw data, which in your case are radiographies.

Yes, but now we have a different problem. Learning machines are very powerful machines, but whenever you read articles on new applications you find that a good chunk of the effort is now in the construction of the dataset. As they say, *"silly data leads to silly visualization"* or, in our case, to silly predictions. So now, instead of spending a lot of time hand-crafting the preprocessing phase, we spend a lot of time crafting the training dataset. You know the drill, unbalanced data, or datasets not representing the actual distribution of an illness in the population, or not taking into account population subgroups. All of that can still lead to a lack of generality, and the predictions translate badly to different scenarios.

There has been this algorithmic advancement that certainly covered a part of the work that was tiring, a lot of preprocessing went away, and so forth. However, these technologies, although very powerful, still have a strong dependence

on the dataset, both on the quality of the data and on the quantity. Do you think things will change from this point of view, too?

Thank goodness the idea of standardization, i.e. preparing data in a machine-understandable standard, is now spreading a bit. I started with the first young doctors at the Milan Polyclinic who began to believe in the applicability of machine-learning and who asked me if I could prepare predictive algorithms for specific applications. And yet, when I received the datasets, they were Excel®files with a maximum of one hundred patients; and sometimes, in cells that were supposed to contain a single number, I found what was a small assay. <smile>

Things have certainly changed: now, working with UC Berkley and the Jackson Laboratory, within the context of the N3C platform provided by the NIH (National Institutes of Health, US), I have had access to data on 14 million patients. One can easily get excited: you think you'll get all this data and you will make all kinds of predictions. Then it happens that if you want to understand if some drugs can be useful in fighting COVID-19, for example, you have to identify a cohort of patients who have certain characteristics. You cannot include patients who perhaps also take another drug that has the opposite effect and inhibits the result of the one you want to test. So, you need to identify a set of patients that is consistent with the problem you want to address. I'm talking about the medical field, but it will be like this in all fields. Data curation is still essential, and I don't see it going away in the near future.

Could we say that data curation is becoming more important than the choice of the specific algorithm?

First of all, what kind algorithms are we talking about? Because on the scene the most used are currently Large (Natural) Language Models (LLMs), the hot topic of the moment. So, if I have a very large dataset, I can try to use those. The problems with these models is the computational cost. You need a lot of money and resources, and environmentally conscious people are already complaining about the amount of energy needed to make LLMs run.

We used 20,000 machine hours on multiple cores at CINECA (Italian supercomputer facility) in a jiffy. And that was just to process highly dimensional data, and even then, it was not enough. Keep in mind that we simply used dimensionality reduction and random forest methods, i.e. a radiomics approach applied to 'omics data. There were no radiological images to begin with, there were only multiomics data. It's really easy to consume computing resources.

And with LLMs you need a great many more hours for training. On one hand, then, we have Language Models that are probably fantastic, but difficult to apply, because you need time, machines, computational power and a gigantic dataset. Alternatively, we have the usual deep learning methods that are now evolving towards graph networks. Graphs are emerging as new computational and data structures that are suitable for representing knowledge, e.g. different types of knowledge, for putting together different sources of information. To answer your question, the only circumstance under which the choice of algorithm doesn't really matter is when you have a lot of money to spare.

Tell us more about the use of graphs as data representation structures

There's this general trend of compressing data using graphs. One example is genomic data based on **coalescent theory** (i.e. tree sequences [137]), which is something we are also working on. One of our ideas is to use graphs as input for neural networks (**graph neural networks**, GNN), thus changing the interface between the data and the neural network.

To do this, you have to build some graphs that reflect the interactions actually existing in nature—for example gene-gene interaction networks, or interaction between proteins or ontologies, things like GO (Gene Ontologies) terms, or Kegg metabolic pathways, or even information regarding RNAs. These knowledge graphs are built in such a way that doesn't refer to individual samples, for example patients, but are rather knowledge graphs that express the biological, or biomedical, knowledge that is known to exist.

This is not something intrinsically new. In fact, it started well before deep neural networks. But they used to be separated realms. You had people working on GO (gene ontologies), people working on protein-protein interaction, or on protein functionalities, and nobody was looking at the other groups. Now, instead, we start from a big collection of information about an entity—it can be a gene, a protein, or something else. For my group, it is usually a gene, and we know when it is expressed, what protein(s) it encodes, what kind of miRNAs are around when the gene is expressed. And all this biomedical knowledge can now be integrated with graphs. It can be integrated to the point that we can, for example, draw the connection between a protein and a drug—a connection usually mediated by the information on a known and well-studied disease. So now, when a new disease appears, we can go and play with these graph representations, and maybe find some candidate drugs among the already existing ones right away.

These graph representation learning methods are able to express all these nodes as vectors, which can then be fed to other models, from other networks to even logistic regression. It's a form of embedding, and if done well I can, for example, combine two nodes to represent the interaction between the two elements and predict whether a link exists between those two nodes. And the two nodes can be, again, a protein, or a new disease, and an existing drug and they can do in-silico drug repurposing. There are studies already published that have done this. For example, for Covid-19, a knowledge graph of this type called KG-Covid was created by our colleagues at Berkeley. They have applied a representation learning graph method to identify possible drugs that could cure Covid [138].

To wrap up, we have a mixture of four models that are bound to appear over and over: (i) natural language processing; (ii) graph representation learning and graph analysis methods (e.g. GNNs); (iii) general deep learning techniques, which at this point are here to stay; and (iv) random forests and logistic regressions.

The last ones are actually very important in the medical field where you often have a few patients each with many features coming from all sorts of analyses, and explainability is really important. Normally, physicians can explain their decisions, why they think a patient is healthy or not. Traditional approaches like random forest models are pretty good at this: it is possible for a human to understand why the model arrived at a certain solution. Obviously, possible doesn't mean easy; it is actually an active research topic, because if for each dataset the model ends up selecting thousands of decision trees, it quickly becomes complicated. But let's say that, in general, it's a type of model that is more similar to a human explicitly reasoning about the data. The same goes for logistic regression, even if here you need to pay attention to the number of features: if there's too many of them, the risk is to just split the importance more or less equally (i.e. penalised regression that leads to the shrinkage of the estimated coefficients). However, the bottom line is that it is doable: you can have methods for selecting the most important features and explain the model.

What are the main limitations of deep learning in the field of life science? And how will this technology drive development in the field of bioinformatics in the coming years?

In my experience, the biggest problem is the lack of open data. For other fields it's easy, you go to *wikidata* (https://en.wikipedia.org/wiki/Wikidata), it's all free. For us, it's more difficult, you need to respect the patient's privacy, there's a lot of rules. I understand that the laws are there for a reason, but sometimes it really slows down the process; and in general it would be nice if there were more open datasets. For example, N3C (the National COVID Cohort Collaborative initiative) has made this huge effort to build a database of more than 14 million patients. When I saw it, I almost cried; you understand the sheer magnitude of possibilities. And even if COVID-19 is over, there's still a lot of potential there, a lot of useful information.

Another big repository is the ALL OF US project, in the United States (https://en.wikipedia.org/wiki/All_of_Us_(initiative)). They are collecting the genomic data of many different individuals, which is particularly useful for rare diseases—which are less and less rare, by the way. To fight these—and similar—battles, we need data computational power. Luckily, the field is moving toward open solutions, so to publish you need to release most of the data used for the experiments.

Apart from data, we want to further investigate the creation and analysis of knowledge graphs and ensure that they are integrated with interaction graphs or dynamic graphs. The idea is to insert patient data from a precise snapshot in time and model the changes over time, carrying out the dynamic evolution of the graphs in order to then identify biomarkers.

The idea is to build a sort of artificial expert that integrates both specific knowledge of the patient and general knowledge of the field, right?

That's right. The classical machine-learning approach is to have all information from a patient, their description, blood values, transcriptomics, everything. And the patient is classified based on that. However, using knowledge graphs as we do, you can also add the reasoning information, for example, on proteins and genes. That information is general to all humans, and is thus valid for everybody. This is an extra layer of reasoning and information manipulation compared to classical machine-learning. This is the frontier of personalized medicine.

So far, you talked about extending current trends. Let's push our predictions and talk about the medium-long range. In five, ten, twenty years, what is going to change? What is your dream application?

As far as I can tell, it's all going to be about graphs and their interactions. Static graphs will encode general knowledge, and dynamic graphs will encode the specific patient history and evolution. And since these graphs are becoming larger and larger, we'll use more and more applications like Natural Language Models.

No, about my dream application: well, my path is mainly about fighting cancer and rare diseases, so I would like an application that extracts biomarkers for curing any kind of disease. Ideally, these predictions should then be easily used through ad hoc blood analysis or other cheap, non-invasive exams. And prognoses, it would be great to create the conditions for the hospitals to deliver precise prognoses.

How will the jobs—as scientists, but also as physicians—be impacted?

Physicians today already want to understand more and more about what machine-learning is. I know of doctors who have at least some R capabilities, they do some analysis by themselves. I think that bioinformaticians and statisticians will be seen more and more as advisers. We'll help them avoid pitfalls and technical difficulties but, ultimately, it will be the doctors who will run the models that help with decision-making. This is very important, because, as you well know, statistical models will always produce an output, and if you misuse the models, the output may well be an hallucination: we actually talk of **hallucinating models**. As a matter of fact, this scares me a little. Sometimes you see published papers that make use of sophisticated methods, and none of the authors have informatics training. We go back to the black-box, the model (or the algorithm, the software package) is trusted without question. That's why I really think that integrating our two professions, physician and machine-learning expert, is of paramount importance.

Do you see any risk of abuse?

Well, yes, any system can be abused. Actually, natural language models are particularly exposed to this risk. You can always ask ChatGPT or some other tool to help you with your task, and if your task is to diagnose an illness, the matter becomes very serious very quickly. The problem is that as of today (June 2023), language models give answers that sound believable but, if you scratch the surface a bit, don't hold. But you need to already be an expert to understand where they actually fail. So, it's good if you use them to help you write—here the emphasis is on the fact that it's actually YOU having control and making the big decisions. But if you just have the models taking the reins, this is bound to hurt somebody badly.

Graph Neural Networks (and Knowledge Graphs)

Graph neural networks (**GNNs**) are networks that take as input **graph data**. Graph data have a form that is actually very similar to the internal structure

(continued)

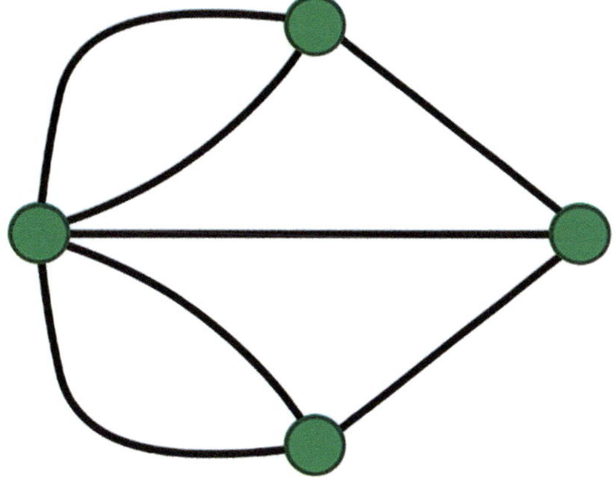

Fig. 7.5 Graph representation of the famous 7 bridges of Koenigsberg: the green dots are the nodes of the graph—representing land masses—while the connections (edges) represent the seven bridges

of a fully-connected neural network: nodes (units) and edges (connections). Figure 7.5 gives the graph representation of the famous problem of the seven bridges of Königsberg (https://en.wikipedia.org/wiki/Seven_Bridges_of_K%C3%B6nigsberg): this abstract representation was formulated by Leonhard Euler, and is considered the foundation of graph theory.

In principle, graphs can be represented as tables in a relational database (RDB), and the other way around. However, for more complex graph structures, such as directed or weighted graphs, the RDB representation may easily become too cumbersome and impractical. In addition, it may be easier and more straightforward to query a graph using specialized query languages—like CYPHER (https://neo4j.com/docs/cypher-manual/current/introduction/)—rather than SQL. The greatest advantage of graph databases is their ability to efficiently represent, query, and analyze complex relationships in the data, making them particularly well-suited for use cases such as social networks, recommendation systems, fraud detection, network and IT operations, and knowledge graphs.

An example of data that can be conveniently represented as a graph is molecules: molecules are made of atoms connected by chemical bonds, and it is easy to think of atoms as the nodes of the graph (with maybe their atomic number as the value) and of chemical bonds as the connections (the length of the connection will give the distance between the atoms in the molecule). Graph data are typically **non-Euclidean data**, i.e. the distance between two data-points is not necessarily a straight line—like on the surface of a sphere—and the relative position of unconnected nodes is irrelevant (Euclidean data are instead representable in a multidimensional linear space, as the multidimensional arrays/tensors in linear algebra).

(continued)

Graph data can be used to predict: (i) properties of the entire graph (e.g. the smell of a molecule given its structure—analogous to image classification); (ii) properties of individual nodes (e.g. based on their connection, which are the atoms that constitute the molecule—analogous to image segmentation); (iii) properties of the edges (e.g. the type of chemical bonds—covalent, double, etc.—based on the relative positions of the atoms).

Graphs are invariant to node ordering, and this makes it challenging for the types of neural networks examined so far to process them: both CNN and RNN assume that the order of pixels or elements in the sequence matters. Additionally, CNN, RNN and fully-connected NNs assume that examples (observations) are independent of each other, but this does not hold for graphs in which each node is linked to others through various types of connections (this is actually a nice property that can help model data dependencies with neural networks).

A graph can enter a neural network in many possible ways (many possible **graph-data representations**): a (relatively) simple approach is to decompose the graph into its **structure** and **signals**. The structure is basically given by the connections between nodes, which can be represented as an **adjacency matrix A**: this is a $(n \times n)$ square matrix that tells who is connected to who. For the 4 nodes in the graph in Fig. 7.5 (from left to right), we can have the adjacency matrix in Eq. 7.1 (the integers represent the number of pairwise connections between nodes):

$$\mathbf{A} = \begin{bmatrix} 0 & 2 & 2 & 1 \\ 2 & 0 & 0 & 1 \\ 2 & 0 & 0 & 1 \\ 1 & 1 & 1 & 0 \end{bmatrix} \tag{7.1}$$

The signal is given by the vector of feature values for each node: for the example of the Königsberg's bridges, these could be the size of the neighbourhoods in km^2, their population, their average per-capita income. This representation of the graph data can be fed as input into a GNN, typically through an encoder: the GNN then will have a specific architecture depending on the task (e.g. convolutions for image data, a final sigmoid or softmax output layer for whole-graph classification, etc.).

If we think a bit harder, we realise that images and text/sound data are just special cases of the more general family of graph data: images can be thought of as fixed-size grid graphs, while text and speech are sequences and can be thought of as (ordered) line graphs. GNNs are linked to the concept of **geometric deep learning** [139], in which geometric priors (constraints) play a fundamental role: to provide some intuition, when we consider an image as

(continued)

a 2-dimensional structure rather than a d-dimensional vector of pixels, we are imposing a geometric constraint on our problem. Such constraints allow us to restrict the number of functions that can be learned with the input data and make the high-dimensional problem tractable (thereby dodging the perilous 'curse of dimensionality').

GNNs are a very broad subject [140, 141], with many applications (e.g. very recently, the discovery of a structural class of antibiotics [142]), and this box aims only to give you the gist of it.

7.5 Natural Language Models

We asked a series of standard questions, some similar to those that we asked the human experts (see above), using three of the most popular large language models (LLMs) freely available on-line: **ChatGPT 3.5** (OpenAI), **Gemini** (previously Bard, Google), **Bing** (Microsoft). These were general questions on future scenarios and the overall impact of deep learning: the answers provided by the models were pretty generic and not too dissimilar. Therefore we decided to report a synoptic summary table of these answers below (Table 7.1), while the full-length answers can be found in Page A.17. In the next section we will investigate how these LLMs perform when asked more technical questions.

How do LLMs Perform on Technical Questions?

Above (Sect. 7.5) we asked popular LLMs a number of generic questions on deep learning, its impact and its future. The answers were not dissimilar to a sort of aimless rambling: not very interesting. Therefore we summarised them in a synoptic table and moved the full-length answers to the Appendix. We then tried to ask the same LLMs more technical questions, specifically those arising from the interviews with the human experts (see Chap. 7, the highlighted boxes).

With this better defined and narrower task the LLMs performed decisevely better: we can say that ChatGPT 3.5. gave by far the more coherent, informative and interesting answers, with Gemini and Bing lagging somewhat behind.

Q1: In machine-learning, what are the possible solutions to the problem of out-of-distribution samples? Use maximum 100 words. Do not use bullet points or lists.

ChatGPT 3.5: To address out-of-distribution samples in machine-learning, one can use techniques like adversarial training, data augmentation, and outlier detection. Adversarial training enhances model robustness by introducing adversarial

Table 7.1 Q&A synoptic table. Maximum 40-word summaries of answers provided to A.I. to our research questions. Each answer was summarised by the same language model that provided the answer

What are the advantages and limitations of using deep neural network models compared to classical machine-learning methods?		
ChatGPT 3.5: Deep neural networks (DNNs) offer automated feature learning and handle complex data relationships, ideal for big data tasks. However, they demand ample labeled data, computational resources, and suffer from interpretability issues and susceptibility to overfitting and adversarial attacks.	**Gemini**: Deep neural networks shine in pattern recognition but require vast data and lack interpretability. Classical ML is interpretable and data-efficient but struggles with complex tasks. Choose the tool that best fits your data, accuracy needs, and interpretability requirements.	**Bing**: Deep Neural Networks (DNNs) excel in complex pattern recognition but require large data and are hard to interpret. Classical machine-learning methods are easier to interpret, work well with smaller datasets, but may struggle with complex data. The choice depends on the task, data nature, resources, and interpretability importance.
What are the ethical implications of the recent advancements on deep learning?		
ChatGPT 3.5: Advancements in deep learning raise ethical concerns: biased decision-making from biased datasets, privacy breaches with sensitive data, lack of transparency, job displacement, and misuse risks. Proactive ethical guidelines are crucial for responsible deployment.	**Gemini**: Deep learning's progress sparks ethical concerns: bias, privacy, job displacement, and societal impact. We need diverse data, transparency, regulations, and collaboration to ensure responsible development for a better future.	**Bing**: Deep learning advancements raise ethical concerns, including algorithmic bias, privacy issues due to extensive data requirements, and decision-making transparency. These technologies can exacerbate societal issues, such as unequal healthcare provision, and require careful ethical considerations.
What will be the Impact of deep learning on life science professions?		
ChatGPT 3.5: Deep learning revolutionizes life sciences: drug discovery, personalized medicine, genomics, biomedical imaging, and healthcare delivery. Improves efficiency in drug development, enables tailored treatments, enhances disease diagnostics, and optimizes healthcare delivery. Challenges include data quality and ethical considerations.	**Gemini**: Deep learning is transforming life sciences: from drug discovery to robotic surgery, it's revolutionizing research, diagnosis, and treatment. Life scientists will need to adapt and collaborate with AI specialists to navigate ethical considerations and unlock the full potential for a healthier future.	**Bing**: Deep learning is transforming life sciences by accelerating drug development, optimizing clinical trials, improving supply chain logistics, and providing insights into biological data. Its impact spans from genomics to biophysics, shaping the future of the industry.

(continued)

Table 7.1 (continued)

What is the future of deep learning development?		
ChatGPT 3.5: Future of deep learning: Architectural innovations, continual learning, interdisciplinary applications, self-supervised learning, ethical AI, federated learning, explainability, hardware acceleration. Emphasis on innovation, collaboration, and ethical development for societal impact.	**Gemini**: Deeper, more efficient AI with explainability and lifelong learning are on the horizon. Bridging the gap with neuroscience and specializing models hold promise. Collaboration and ethical considerations are crucial for a positive future powered by deep learning.	**Bing**: Deep learning, a leading AI technology, is expected to evolve further, despite differing opinions on its potential. Its ability to improve prediction accuracy and learn from unstructured data has led to operational and financial benefits. New research areas like few-shot learning and GANs are strengthening deep learning models, with applications across AI.

examples during training.[4] Data augmentation involves generating diverse data to better generalize the model. Outlier detection identifies and handles data that significantly deviates from the training distribution. Additionally, utilizing ensemble methods and uncertainty estimation can improve model predictions for unfamiliar samples.

Gemini: machine-learning models can be fooled by data unlike their training. To tackle this, researchers explore methods for both identifying these outliers (out-of-distribution detection) and training the main model to express its confidence in predictions. This allows flagging uncertain cases that might be outliers.

Bing: Handling out-of-distribution samples in machine-learning involves several strategies. Anomaly detection models identify anomalies, uncertainty estimation leverages models providing uncertainty estimates (e.g., Bayesian neural networks), outlier rejection sets confidence thresholds, data augmentation enhances training data, domain adaptation fine-tunes models on in-distribution samples, and ensemble methods combine predictions for robustness. Remember, no single solution fits all scenarios!

Q2: Can you describe the basics of transformers in deep learning? Use maximum 100 words. Do not use bullet points or lists.

ChatGPT 3.5: Transformers in deep learning are neural network architectures designed for sequence-to-sequence tasks, excelling in natural language processing. They rely on self-attention mechanisms[5] to weigh the importance of different

[4] (Authors' note) Adversarial examples are examples whose features are modified so to induce incorrect classification (optical illusions for DL algorithms). For instance, a spam email message can be modified by mispelling "bad" words and adding "good" words. Actually, feature modifications are usually very small, almost imperceptible for humans, but sufficient to deceive the machine.

[5] For a review on self-attention in DL you can read "Attention, please! A survey of neural attention models in deep learning" by de Santana Correia & Colombini, 2022 https://link.springer.com/article/10.1007/s10462-022-10148-x.

words in a sentence, capturing context and dependencies regardless of distance. This enables parallel processing of sequences, unlike recurrent neural networks. Transformers consist of encoder and decoder layers with multi-head attention, feed-forward neural networks, and normalization techniques. Their ability to model complex relationships and process long-range dependencies efficiently has led to breakthroughs in machine translation, text generation, and other NLP applications.

Gemini: Transformers are a deep learning architecture for sequences like text or speech. Unlike older models, they can analyze all parts of the sequence at once, thanks to a mechanism called self-attention. This lets them understand relationships between words or sounds better, leading to improved tasks like machine translation and text summarization.

Bing: Transformers are a groundbreaking neural network architecture that revolutionized natural language processing (NLP) and other sequence-to-sequence tasks. Key features include self-attention, an encoder-decoder architecture, positional encoding, multi-head attention, feed-forward neural networks, and layer stacking. They excel in tasks like machine translation and text summarization, powering state-of-the-art NLP models like BERT, GPT, and T5.

Q3: Can you explain the concept of attention in deep learning? Use maximum 100 words. Do not use bullet points or lists

ChatGPT 3.5: Attention in deep learning is a mechanism that dynamically focuses on relevant parts of the input data when making predictions. It assigns varying importance weights to different elements, allowing the model to prioritize significant information while ignoring irrelevant details. This is crucial in tasks like language translation, where understanding context and relationships between words is essential. By enabling models to attend to specific parts of the input, attention improves the handling of long-range dependencies and enhances performance in complex sequence tasks, forming the foundation of advanced architectures like transformers.

Gemini: In deep learning, attention mimics human focus. It allows models to prioritize relevant parts of an input, like specific words in a sentence. The model assigns weights to different elements, focusing on the most important ones. This improves tasks like translation by letting the model concentrate on crucial parts of the input sequence when generating the output.

Bing: Attention is a pivotal concept in deep learning, especially in natural language processing (NLP) and sequence-to-sequence tasks. It enables models to focus on relevant parts of input data during predictions or output generation. Key components include self-attention, an encoder-decoder architecture, positional encoding, multi-head attention, feed-forward neural networks, and layer stacking. Transformers, powered by attention, excel in tasks like machine translation and text summarization.

Q4: In statistics, what is active learning? Use maximum 100 words. Do not use bullet points or lists.

ChatGPT 3.5: Active learning in statistics is a technique where the model selectively queries the most informative data points from a pool of unlabeled data for labeling. This approach aims to improve model performance efficiently by focusing on data that is expected to enhance learning the most. It is particularly useful when labeled data is scarce or expensive to obtain. By iteratively selecting and labeling the most valuable samples, active learning reduces the amount of labeled data required while maintaining or improving the model's accuracy, making it a cost-effective strategy for training machine-learning models.

Gemini: Active learning in statistics, also called optimal experimental design, is an approach where the statistical model itself drives data collection. Instead of passively using a pre-collected dataset, the model chooses which new data points would be most informative to learn from. This can be especially useful when acquiring data is expensive or time-consuming.

Bing: Active learning is a method where a learning algorithm interacts with a human user (or another information source) to label new data points with desired outputs. It constructs valid confidence intervals and hypothesis tests while leveraging any black-box machine-learning model and handling any data distribution. Active learning achieves the same accuracy with fewer samples than non-adaptive baselines, making it valuable for efficient learning.

Q5: Can you describe the basics of graph neural networks? Use maximum 100 words. Do not use bullet points or lists.

ChatGPT 3.5: Graph neural networks (GNNs) are a class of neural networks designed to operate on graph structures, where data is represented as nodes connected by edges. GNNs iteratively update node representations by aggregating information from their neighbors using message-passing mechanisms. This allows the network to capture the dependencies and relationships within the graph. By combining local and global information, GNNs excel in tasks like node classification, link prediction, and graph classification. Their ability to learn from the complex topology of graphs makes them powerful for applications in social networks, molecular biology, and recommendation systems.

Gemini: Imagine a neural network that can analyze graphs, like social media connections or molecules. Graph neural networks (GNNs) are special deep learning models for this. Unlike image recognition models that process pixels in a grid, GNNs focus on relationships between nodes (data points) in a graph. They iteratively update each node's information by considering its own features and those of its connected nodes. This allows them to learn patterns from the graph structure itself, useful for tasks like node prediction or link forecasting.

Bing: Graph Neural Networks (GNNs) are a class of neural network architectures designed for processing graph-structured data. They operate on graphs where nodes represent entities (e.g., users, molecules), and edges represent relationships (e.g., social connections, chemical bonds). GNNs learn embeddings for each node by aggregating information from neighboring nodes using techniques like message passing and graph convolutions. These models stack multiple layers to capture increasingly complex graph patterns. GNNs excel in tasks like node classification, link prediction, and graph classification, bridging the gap between neural networks and structured graphs

Python Notebooks

A

All the notebooks are thought of as teaching exercises. We favoured code simplicity and clarity over optimization. You can just take a look at the code and results or import the notebooks into your preferred environment to run and tweak them. All notebooks are hosted long-term in the following Github repository https://github.com/ne1s0n/bioinformateachers/blob/main/dlb/.

The specific notebook names are provided below, although we invite you to keep in mind that the above Github folder is obviously dynamic and may host new notebooks over time.

A.1 Notebook—K-Means

In this notebook, we implement from scratch the K-means clustering algorithm and use it to color a geographical map. In doing so, we show the key concepts related to the algorithm, plus some ideas common to all clustering problems. You can find the notebook at: https://github.com/ne1s0n/bioinformateachers/blob/main/dlb/archipelago/lab_full.ipynb.

A.2 Notebook—Confusion Matrix

In this notebook, we illustrate how to construct, read and interpret a confusion matrix from classification problems. Binary and multiclass classification problems will be addressed, to show the different types of errors that can be made. Summary metrics for the different error types are presented: e.g. true/false positive/negative rates (TPR, FPR, TNR, FNR), positive/negative predictive values, false discovery rate (FDR).

You can find the notebook at: https://github.com/ne1s0n/bioinformateachers/blob/main/dlb/measure_model_performance/confusion_matrix.ipynb.

F. Biscarini, N. Nazzicari, *Deep Learning for Life Sciences*, Decoding Evolution 1,
https://doi.org/10.1007/978-3-031-96852-5

A.3 Notebook—ROC, AUC, MCC

In this notebook we build upon the previous notebook on confusion matrices, and show how to summarise the different types of errors into single summary metrics that will help you to evaluate correctly the model's performance and compare different models: ROC curves, AUC (area under the curve), MCC (Matthews correlation coefficient).

You can find the notebook at: https://github.com/ne1s0n/bioinformateachers/blob/main/dlb/measure_model_performance/roc_auc_mcc.ipynb.

A.4 Notebook—Additional Metrics

In this notebook we introduce additional model performance metrics that can be used for classification and regression problems: Cohen's kappa, calibration plots, Brier score, normalized discounted cumulative gain (NDCG).

You can find the notebook at: https://github.com/ne1s0n/bioinformateachers/blob/main/dlb/measure_model_performance/additional_metrics.ipynb.

A.5 Notebook—Backpropagation

With a numerical example, we illustrate how forward- and backpropagation work to learn the weights (parameters) of a neural network.

You can find the notebook at: https://github.com/ne1s0n/bioinformateachers/blob/main/dlb/backpropagation/backpropagation.ipynb.

A.6 Notebook—Function Approximation

In this notebook, we start with a simple example—a parabola—and show how a neural network can approximate it. We also show how data normalization improves the network's performances, and what happens if you choose higher-degree polynomials.

You can find the notebook at: https://github.com/ne1s0n/bioinformateachers/blob/main/dlb/function_approximation/function_approximation.ipynb.

A.7 Notebook—Logistic Regression

In this notebook, we illustrate how to perform logistic regression using a neural network-like implementation with one single unit (perceptron). This is equivalent to a logistic regression model, only it is solved using neural networks (forward and backpropagation) rather than with maximum likelihood (or similar algorithms).

You can find the notebook at: https://github.com/ne1s0n/bioinformateachers/blob/main/dlb/regression/logistic_regression.ipynb.

A.8 Notebook—Softmax

In this notebook, we use a very simple one-layer neural network to extend logistic regression to multiclass classification using softmax regression. This is equivalent to a multinomial logistic regression model, but it is solved using neural networks (forward and backpropagation) rather than maximum likelihood (or similar algorithms). You can find the notebook at: https://github.com/ne1s0n/bioinformateachers/blob/main/dlb/regression/softmax_regression.ipynb.

A.9 Notebook—Linear Regression with NN

In this notebook, we show how you can use a simple single-layer neural network to solve a linear regression problem. This is equivalent to a common linear regression model, only it is solved using neural networks (forward and backpropagation) rather than with ordinary least squares (OLS) or maximum likelihood (or similar algorithms).

You can find the notebook at: https://github.com/ne1s0n/bioinformateachers/blob/main/dlb/regression/linear_regression.ipynb.

A.10 Notebook—From Basic Regression to Deeper NN Models

We extend the material presented in previous notebooks to illustrate how you can move from simple single-layer neural networks to "real" neural network models with multiple nodes and multiple layers.

You can find the notebook at: https://github.com/ne1s0n/bioinformateachers/blob/main/dlb/dense_NN/basic_dense_NN_models.ipynb.

A.11 Notebook—Basic RNN

We'll use clear numerical examples to guide you through some simple RNN (recurrent neural networks) models: the objectives are to understand the input data type and shape, to calculate the number of trainable parameters involved, and to come to know the basic architecture of RNNs. The way in which to implement RNNs in Python is also discussed.

You can find the notebook at: https://github.com/ne1s0n/bioinformateachers/blob/main/dlb/RNN/basic_RNN_models.ipynb.

A.12 Notebook—Specialized RNN

Basic RNN models typically suffer from the curse of vanishing gradients. In this notebook, we extend the basic RNN model and show more advanced architectures for analyzing sequence (longitudinal) data: GRU (gate-recurrent unit) and LSTM (long-short term memory) units/layers.

You can find the notebook at: https://github.com/ne1s0n/bioinformateachers/blob/main/dlb/RNN/advanced_RNN_models.ipynb.

A.13 Notebook—Data Augmentation

This notebook illustrates the principles and practical implementation of data augmentation, and shows how this data manipulation technique can improve the performance of CNN models for image recognition.

You can find the notebook at: https://github.com/ne1s0n/bioinformateachers/blob/main/dlb/data_augmentation/data_augmentation.ipynb.

A.14 Notebook—Embeddings

In this notebook we provide some intuition behind the fundamental concept of data embeddings, which is an alternative and very effective way of representing the input data.

You can find the notebook at: https://github.com/ne1s0n/bioinformateachers/blob/main/dlb/embeddings/embeddings.ipynb.

A.15 Notebook—Transfer Learning

In this notebook, we introduce the basic concepts of transfer learning, and how it can be done with large neural networks in Python.

You can find the notebook at: https://github.com/ne1s0n/bioinformateachers/blob/main/dlb/transfer_learning/transfer_learning.ipynb.

A.16 Notebook—Feature Selection

In this notebook we'll use a breast cancer dataset to investigate if neural networks can be used for feature selection. We'll use the learned weights as a proxy for feature importance, and see what happens with a super-simple one-layer network and how things get complicated when we add just one more layer. You can find the notebook at: https://github.com/ne1s0n/bioinformateachers/blob/main/dlb/inside_the_network/feature_selection.ipynb.

A.17 Notebook—Opening the Black Box

This notebook tries to crack the neural network black box. It is explored the concept of variable importance from NN models: how is it possible to extract relevant variables from the network? Approaches based on permutations, reverse engineering, attribution matrices and relevant image pixels will be explored.

You can find the notebook at: https://github.com/ne1s0n/bioinformateachers/blob/main/dlb/inside_the_network/opening_black_box.ipynb.

Natural Language Models Interviews

B

Below is the full-length text of the answers given by the tested natural language models to a series of standard questions, some similar to those we asked the human experts in Chap. 7. All the text is copied as-is and thus contains the occasional grammatical error or nonsensical bit. Developing an eye for spotting these kinds of details is quickly becoming a necessary skill for any good scientist (or, to be completely frank, any informed citizen).

What are the advantages and limitations of using deep neural network models compared to classical machine-learning methods? Don't use bullet lists

ChatGPT 3.5 (OpenAI): Deep neural network (DNN) models offer several advantages compared to classical machine learning methods. They automatically learn features from raw data, eliminating the need for manual feature engineering. Additionally, they can model complex, non-linear relationships in data, making them suitable for capturing intricate patterns that classical methods might struggle with. DNNs are highly scalable and can handle large datasets, making them suitable for big data applications. Their hierarchical representation enables better generalization to unseen data, and they can perform end-to-end learning, bypassing the need for intermediate processing steps.

However, DNNs also come with limitations. They require large amounts of labeled data for training, which can be a challenge in domains where labeled data is scarce. Training DNNs is computationally intensive and often requires specialized hardware. DNNs are often regarded as black box models, making it difficult to interpret their decisions. They are prone to overfitting, especially when the model capacity is high relative to the size of the training data, requiring careful regularization. Tuning the many hyperparameters of DNNs can be time-consuming. Moreover, DNNs have been shown to be vulnerable to adversarial attacks, where small perturbations to the input can cause the model to make incorrect predictions. Therefore, while DNNs have revolutionized many fields, they should be used judiciously, considering the specific requirements and constraints of the problem at hand.

F. Biscarini, N. Nazzicari, *Deep Learning for Life Sciences*, Decoding Evolution 1,
https://doi.org/10.1007/978-3-031-96852-5

Gemini (Previously Bard, Google) Deep neural networks (DNNs) and classical machine learning (ML) methods are powerhouses in the field of AI, each excelling in different aspects and facing unique challenges. Understanding their strengths and weaknesses helps us choose the right tool for the job.

DNNs: The Powerhouse of Pattern Recognition. Advantages:

- Unveiling hidden patterns: DNNs excel at identifying complex, non-linear relationships in massive datasets, making them ideal for tasks like image recognition, natural language processing, and speech recognition.
- Automatic feature engineering: Unlike classical ML, DNNs automatically learn features from the data, saving time and expertise in feature selection.
- State-of-the-art performance: In specific domains like computer vision and language understanding, DNNs often achieve unmatched accuracy, pushing the boundaries of what's possible.

Limitations:

- Data dependency: DNNs are data hungry, requiring vast amounts of training data to perform well. This can be a hurdle for applications with limited data availability.
- Black box nature: DNNs' decision-making process can be opaque, making it difficult to understand why they make certain predictions. This lack of interpretability can raise concerns in sensitive applications.
- Computational cost: Training DNNs requires significant computational power, making them expensive and resource-intensive to run.

Classical ML: The Reliable Workhorse Advantages:

- Interpretability: Classical ML models are often easier to understand, allowing for clearer insights into their decision-making process. This is crucial for applications requiring explainability and fairness.
- Data efficiency: Many classical ML algorithms perform well with smaller datasets, making them suitable for scenarios where data is limited.
- Lower computational cost: Training and running classical ML models generally require less computational power compared to DNNs.

Limitations:

- Performance complexity: For tasks involving intricate patterns and relationships, classical ML models may struggle to match the accuracy achieved by DNNs.
- Feature engineering burden: Extracting relevant features from data often requires manual effort and expertise, which can be time-consuming and domain-specific.

Choosing the Right Tool: The best choice between DNNs and classical ML depends on the specific problem at hand. Consider factors like data availability, desired accuracy, need for interpretability, and computational resources. In some cases, hybrid approaches combining both techniques can leverage their individual strengths for optimal results.

Remember, both DNNs and classical ML have their place in the AI landscape. Understanding their distinct characteristics empowers us to make informed decisions and unlock the full potential of these powerful tools.

Bing Notebook (Microsoft) Deep Neural Networks (DNNs) and classical machine learning methods each have their unique advantages and limitations.

DNNs are powerful tools for modeling complex patterns and structures in data. They excel in tasks where the data has a hierarchical or spatial structure, such as image and speech recognition. DNNs can automatically learn and extract features from raw data, eliminating the need for manual feature engineering which is often required in classical machine learning methods. Furthermore, DNNs are highly flexible and can be adapted to a wide range of tasks by adjusting their architecture and training procedures.

However, DNNs also have their limitations. They require large amounts of data and computational resources for training, which can be a barrier for small-scale projects or for use in resource-constrained environments. DNNs are often seen as "black boxes" due to their complex internal workings, making them difficult to interpret and understand. This lack of interpretability can be problematic in fields where understanding the decision-making process is crucial, such as healthcare or finance.

On the other hand, classical machine learning methods, such as decision trees, linear regression, and support vector machines, are generally easier to interpret. These methods often require less computational resources and can work well with smaller datasets. They are also grounded in well-established statistical theory, providing a level of robustness and reliability.

However, classical methods may struggle with complex, high-dimensional data, and often require careful feature engineering and data preprocessing. They may not perform as well as DNNs on tasks involving unstructured data, such as image or speech recognition.

In conclusion, the choice between DNNs and classical machine learning methods depends on the specific task, the nature of the data, the available resources, and the importance of interpretability

What are the ethical implications of the recent advancements on deep learning? Don't use bullet points

ChatGPT 3.5 (OpenAI): Recent advancements in deep learning have raised various ethical implications that need careful consideration. One significant concern is the potential for biased decision-making. Deep learning models learn from large datasets, which may reflect biases present in society, leading to discriminatory outcomes. For example, facial recognition systems trained on biased datasets may exhibit higher error rates for certain demographic groups, leading to unfair treatment.

Privacy is another ethical concern. Deep learning models trained on sensitive data, such as medical records or personal communications, raise questions about data privacy and consent. There's a risk of unauthorized access or misuse of such data, leading to breaches of privacy rights.

Transparency and interpretability are crucial ethical considerations. Deep learning models are often complex and difficult to interpret, making it challenging to understand how they arrive at their decisions. This lack of transparency can erode trust and accountability, particularly in high-stakes applications like healthcare or criminal justice.

Furthermore, the automation of tasks previously performed by humans raises concerns about job displacement and socioeconomic inequality. While deep learning technologies have the potential to improve efficiency and productivity, they may also lead to job loss in certain sectors, exacerbating existing inequalities.

Another ethical consideration is the responsible deployment of AI technologies. There's a need for guidelines and regulations to ensure that deep learning systems are used ethically and responsibly. This includes considerations such as fairness, accountability, transparency, and safety throughout the development and deployment lifecycle.

Finally, there are concerns about the misuse of deep learning technologies for malicious purposes, such as surveillance, manipulation, or misinformation campaigns. As deep learning becomes more powerful and accessible, there's a need for safeguards to prevent its misuse and protect against potential harms.

Overall, while deep learning advancements offer tremendous potential for innovation and progress, it's essential to address the ethical implications proactively to ensure that these technologies are developed and deployed in a manner that respects individual rights, promotes fairness and transparency, and mitigates potential harms.

Gemini (Previously Bard, Google)

Recent advancements in deep learning, while impressive, raise profound ethical questions that demand careful consideration. These concerns can be broadly categorized into three main areas:

1. **Bias and Fairness** Amplifying existing biases: Deep learning models trained on biased data can perpetuate and even amplify those biases in their outputs. This can lead to discriminatory outcomes in areas like loan approvals, hiring decisions, and criminal justice. Lack of explainability: The "black box" nature of many deep learning models makes it difficult to understand how they arrive at their decisions. This lack of transparency can be problematic when the model's decisions have significant consequences for individuals or society. Representation and inclusion: If deep learning models are not trained on diverse and representative datasets, they may fail to generalize effectively to different populations. This can lead to unfair outcomes for groups that are underrepresented in the training data.

2. **Privacy and Security** Data collection and use: Deep learning often requires vast amounts of personal data for training, raising concerns about privacy and data security. Individuals may not be aware of how their data is being used, and there may be insufficient safeguards to protect it from misuse. Deepfakes and misinformation: Deep learning techniques can be used to create highly realistic but fake videos and audio recordings, known as deepfakes. These can be used to spread misinformation, damage reputations, and manipulate public opinion. Algorithmic bias and profiling: Deep learning algorithms can be used to create detailed profiles of individuals, which could be used for discriminatory purposes or to unfairly target them with advertising or surveillance.
3. **Societal Impact and Responsibility** Job displacement: Automation powered by deep learning could lead to significant job displacement in various sectors, raising concerns about unemployment and economic inequality. Autonomous weapons and surveillance: The use of deep learning in autonomous weapons systems and mass surveillance raises ethical concerns about accountability, transparency, and the potential for misuse. Human-machine interactions: As deep learning systems become more sophisticated, questions arise about the nature of human-machine interactions and the potential for these systems to influence or manipulate human behavior.

Addressing these ethical challenges requires a multi-pronged approach. Developers need to be more mindful of potential biases in their models and data. Robust regulations and ethical frameworks are needed to govern data collection, use, and security. Open dialogue and collaboration between technologists, ethicists, policymakers, and the public are crucial to ensure that deep learning benefits all of society and aligns with our shared values.

Remember, responsible development and deployment of deep learning are essential to prevent negative societal impacts and ensure these powerful tools contribute to a just and equitable future.

Bing Notebook (Microsoft) The recent advancements in deep learning have brought about a host of ethical implications. One of the key concerns is the issue of algorithmic bias. The outputs of translation and search engine algorithms are largely perceived as objective, yet frequently encode language in gendered ways(1). Bias has also been reported in algorithmic advertisement, with opportunities for higher-paying jobs and jobs within the field of science and technology advertised to men more often than to women(1).

Another significant ethical concern is the dependency on extensive labeled data and training costs. Deep learning models require large amounts of data for training, which raises questions about privacy and data protection(2). Furthermore, the training process can be expensive in terms of time and computational resources(2).

The rise of nationalism in the Ottoman Empire also poses ethical risks.[1] Algorithms are not ethically neutral and their ubiquitous use in society can lead to new ethical problems(1). For instance, prediction algorithms used to manage the health data of millions of patients in the United States exacerbate existing problems, with white patients given measurably better care than comparably similar, black patients(1).

Lastly, the ethical implications of deep learning extend to the realm of decision-making. As AI takes on a bigger decision-making role in more industries, it's crucial to ensure that these decisions are made in an ethical and fair manner(3). This includes considerations about transparency, accountability, and the potential for misuse of AI technologies(4).

Source: Conversation with Bing, 2/8/2024

1. The ethics of algorithms: key problems and solutions. https://link.springer.com/article/10.1007/s00146-021-01154-8.
2. A Review of Deep Transfer Learning and Recent Advancements—MDPI. https://www.mdpi.com/2227-7080/11/2/40.
3. Ethical concerns mount as AI takes bigger decision-making role. https://news.harvard.edu/gazette/story/2020/10/ethical-concerns-mount-as-ai-takes-bigger-decision-making-role/.
4. The Moral And Ethical Implications Of Artificial Intelligence. https://stefanini.com/en/insights/articles/the-moral-and-ethical-implications-of-artificial-intelligence.
5. undefined. https://doi.org/10.3390/technologies11020040.[2]
6. Getty Images. https://www.gettyimages.com/detail/photo/deep-learning-logo-royalty-free-image/871793108.[3]

What will be the Impact of deep learning on life science professions? Don't use bullet points

ChatGPT 3.5 (OpenAI): The impact of deep learning on life science professions is expected to be profound and far-reaching. Deep learning techniques have the potential to revolutionize various aspects of life sciences, including drug discovery, personalized medicine, genomics, biomedical imaging, and healthcare delivery.

In drug discovery, deep learning models can analyze vast amounts of biological and chemical data to identify potential drug candidates more efficiently. By predicting molecular properties, simulating drug-target interactions, and optimizing drug design, deep learning accelerates the drug development process, reducing costs and increasing success rates.

Personalized medicine stands to benefit significantly from deep learning advancements. By analyzing individual genetic and clinical data, deep learning

[1] This is a good example of the network producing texts which is correct from the grammatical point of view but contains little actual meaning.

[2] Broken link.

[3] Also a broken link.

models can predict disease risk, recommend tailored treatments, and optimize patient outcomes. Deep learning also enables the discovery of biomarkers for early disease detection and monitoring, leading to more precise diagnostics and targeted interventions.

In genomics, deep learning techniques are transforming our understanding of genetic data. Deep learning models can analyze DNA sequences, identify genetic variants associated with diseases, and predict the functional impact of genetic mutations. This knowledge has implications for disease prevention, diagnosis, and treatment, paving the way for personalized genomic medicine.

Biomedical imaging is another area where deep learning is making a significant impact. Deep learning models can analyze medical images such as X-rays, MRIs, and histopathology slides with remarkable accuracy, aiding in disease diagnosis, treatment planning, and prognosis prediction. Deep learning also enables image enhancement, reconstruction, and segmentation, improving the quality and efficiency of medical imaging workflows.

In healthcare delivery, deep learning technologies are being deployed to optimize clinical decision-making, patient monitoring, and resource allocation. Deep learning models can analyze electronic health records, predict patient outcomes, and identify opportunities for intervention, leading to more efficient and personalized healthcare delivery. Additionally, deep learning facilitates the automation of routine administrative tasks, freeing up healthcare professionals to focus on patient care.

Overall, deep learning has the potential to transform life science professions by accelerating scientific discovery, enabling personalized medicine, and improving healthcare delivery. However, realizing this potential requires addressing various challenges, including data quality, model interpretability, regulatory compliance, and ethical considerations. With continued research, collaboration, and innovation, deep learning is poised to drive significant advancements in life sciences, ultimately benefiting patients and society as a whole.

Gemini (Previously Bard, Google) Deep learning is poised to revolutionize life science professions, transforming the way research is conducted, diseases are diagnosed, and treatments are developed. Here's a glimpse into this future:

Enhanced Research Drug discovery: DNNs can analyze massive datasets of molecular structures and biological interactions, accelerating the identification of promising drug candidates and reducing development timelines. Personalized medicine: Deep learning algorithms can analyze patient data (genomics, medical history, lifestyle) to predict disease susceptibility and tailor treatment plans to individual needs. Genomics and proteomics: DNNs can analyze vast genomic and proteomic data to uncover hidden patterns, leading to a deeper understanding of disease mechanisms and biomarkers.

Revolutionized Diagnostics Medical imaging: Deep learning algorithms can analyze medical images (X-rays, MRIs) with superhuman accuracy, aiding in early and accurate diagnosis of various diseases. Pathology: DNNs can analyze digital pathology slides to assist pathologists in detecting cancer and other diseases, improving diagnostic accuracy and efficiency. Non-invasive diagnostics: DNNs can analyze data from non-invasive methods like ECGs and blood tests to identify potential health issues, enabling early intervention.

Transformative Interventions Robotic surgery: Deep learning-powered robots can assist surgeons in complex procedures, improving precision and minimizing risks. Prosthetic limbs and exoskeletons: DNNs can interpret neural signals to control prosthetic limbs and exoskeletons in real-time, offering unprecedented levels of control and functionality. Drug delivery systems: DNNs can optimize drug delivery systems, tailoring dosing and release patterns to individual patients for improved efficacy and reduced side effects.

Changing Roles and New Opportunities Life scientists: Will need to upskill in data science, programming, and deep learning to collaborate effectively with AI specialists and interpret the models' outputs. New professions: Roles like AI specialists, data scientists, and computational biologists will become increasingly crucial in life science research and development. Ethical considerations: Addressing bias, fairness, and privacy concerns in AI-driven healthcare will require collaboration between scientists, ethicists, and policymakers.

The impact of deep learning on life science professions will be profound, demanding adaptation and skill development. However, the potential benefits for scientific discovery, diagnosis, and treatment are vast, holding the promise of a healthier future for all. Remember, deep learning is a tool, and its responsible application requires careful consideration of the ethical and societal implications. By harnessing its power ethically, we can unlock a new era of life science advancements that benefit humankind.

Bing Notebook (Microsoft) Deep learning is poised to have a transformative impact on life science professions. It's already making waves throughout the sciences broadly and the life sciences in particular(3).

One of the key areas where deep learning is making a significant impact is in drug development. By analyzing vast amounts of data, deep learning algorithms can help identify potential drug candidates and predict their effectiveness, significantly speeding up the drug discovery process(1).

In addition, deep learning is revolutionizing the way clinical trials are conducted. These algorithms can analyze patient data to determine the optimal candidates for clinical trials, predict patient responses to various treatments, and monitor patient health during the trial(1).

The supply chain is another area where deep learning is making a difference. By predicting demand and optimizing logistics, these algorithms can help ensure that life-saving drugs and medical supplies reach the patients who need them most(1).

Deep learning is also a powerful tool for surveying and classifying biological data(2). For instance, it can be used to analyze genomic data to understand the genetic basis of various diseases, or to analyze medical images to diagnose diseases(2).

Moreover, deep learning is helping to advance biophysics by enabling researchers to analyze complex biological systems and phenomena(3).

In summary, deep learning is transforming life science professions by enabling more efficient drug development, optimizing clinical trials, improving supply chain logistics, and providing new insights into biological data. This technology is likely to shape the future of the life sciences industry for the better(1).

Source: Conversation with Bing, 2/8/2024

1. Deep Learning for the Life Sciences—SearchWorks catalog. https://searchworks.stanford.edu/view/13180512.
2. AI, machine learning transform life sciences industry | EY—US. https://www.ey.com/en_ca/life-sciences/how-are-ai-and-machine-learning-altering-the-life-sciences-industry.[4]
3. Deep learning for biology—Nature. https://www.nature.com/articles/d41586-018-02174-z.
4. Getty Images. https://www.gettyimages.com/detail/photo/deep-learning-logo-royalty-free-image/871793108.[5]

What is the future of deep learning development? Don't use bullet points

ChatGPT 3.5 (OpenAI): The future of deep learning development is poised to be dynamic and transformative, with continued advancements across various fronts. Here are some key trends and directions likely to shape the future of deep learning:

1. Architectural Innovations: Deep learning architectures will continue to evolve, with researchers exploring novel network architectures, such as transformer-based models, graph neural networks, and attention mechanisms. These architectures aim to improve model efficiency, scalability, and generalization across diverse domains.
2. Continual Learning: There is increasing interest in developing deep learning models capable of continual learning, where they can adapt and learn from new data over time without catastrophic forgetting. Continual learning techniques are essential for real-world applications where data distribution may change or evolve over time.
3. Interdisciplinary Applications: Deep learning will increasingly intersect with other disciplines, such as natural language processing, computer vision, robotics, and healthcare. Interdisciplinary collaborations will drive innovation and enable the development of more powerful and versatile deep learning models with broader applicability.

[4] Broken link.

[5] Broken link.

4. Self-Supervised Learning: Self-supervised learning approaches, where models learn from unlabeled data by predicting missing or corrupted portions, are gaining traction. These techniques have the potential to reduce the reliance on labeled data and improve model generalization across tasks and domains.
5. Ethical and Responsible AI: There will be a growing emphasis on ethical and responsible AI development practices within the deep learning community. This includes considerations such as fairness, transparency, accountability, safety, and privacy throughout the AI lifecycle, from data collection and model training to deployment and monitoring.
6. Federated Learning and Edge Computing: Federated learning, where models are trained across distributed devices without centralized data aggregation, and edge computing, where inference occurs locally on devices, will become more prevalent. These approaches enable privacy-preserving and low-latency deep learning applications in resource-constrained environments.
7. Explainable AI: As deep learning models become increasingly complex, there will be a greater emphasis on developing techniques for model interpretability and explainability. Explainable AI methods aim to enhance trust, transparency, and accountability by providing insights into model predictions and decision-making processes.
8. Hardware Acceleration: Advances in hardware technologies, such as specialized AI chips (e.g., TPUs, GPUs) and neuromorphic computing, will continue to drive the development of more powerful and energy-efficient deep learning systems. These advancements will enable the deployment of deep learning models in edge devices and embedded systems.

Overall, the future of deep learning development is characterized by ongoing innovation, interdisciplinary collaboration, and a focus on ethical and responsible AI. As deep learning continues to mature, it will play an increasingly integral role in addressing complex real-world challenges and driving societal impact across various domains.

Gemini (Previously Bard, Google) Predicting the future with certainty is impossible, but peering into the crystal ball of deep learning development reveals tantalizing possibilities and significant challenges to overcome. Here are some key trends that might shape the future:

Deeper and More Efficient Models: We can expect to see deeper and more complex neural networks emerge, capable of tackling even more intricate problems. However, efficiency will be crucial, as training behemoth models requires immense computational resources. Research into specialized hardware and efficient training algorithms will be paramount.

Explainable and Trustworthy AI: The "black box" nature of deep learning is a major hurdle. Explainable AI (XAI) techniques are on the rise, aiming to shed light on model decisions and build trust in their outputs. This is especially critical in sensitive applications like healthcare and finance.

Lifelong Learning and Adaptation: Current deep learning models are typically static, requiring retraining for new information. The future might see models that continuously learn and adapt, mimicking the human brain's ability to process new information on the fly. This would require breakthroughs in incremental learning and memory management.

Bridging the Gap Between AI and Neuroscience: As we delve deeper into the brain's workings, insights might inspire novel neural network architectures and learning algorithms. This interdisciplinary approach could lead to more efficient and human-like AI.

Emergence of Specialized Deep Learning: Instead of one-size-fits-all models, we might see specialized deep learning algorithms tailored to specific domains, like healthcare, finance, or robotics. This specialization could unlock significant performance gains in niche areas.

Collaboration and Democratization: Deep learning development currently requires significant expertise and resources. Democratizing access through user-friendly tools and fostering collaboration between diverse communities could accelerate innovation and broaden the impact of deep learning.

Ethical Considerations Remain Paramount: As deep learning becomes more powerful, ethical concerns will only intensify. Addressing issues like bias, fairness, privacy, and societal impact will be crucial to ensure responsible development and deployment of this technology.

The future of deep learning is brimming with potential, but challenges abound. By addressing these head-on, we can ensure that deep learning serves as a force for good, empowering scientific discovery, improving healthcare, and shaping a better future for all. Remember, responsible development and careful consideration of ethical implications are paramount in shaping the positive impact of deep learning.

Bing Notebook (Microsoft)

Deep learning, currently the most effective AI technology for numerous applications, is expected to continue evolving. There are differing opinions on how capable deep learning can become. While some researchers believe that all problems could be solved with deep learning, others point to flaws in deep learning where remedies are not clear(1).

With increasing interest in deep learning from the general public as well as developer and research communities, there could be breakthroughs in the field. Experts expect such breakthroughs to come from areas such as capsule networks, deep reinforcement learning, and other approaches that complement deep learning's current limitations(1).

The interest in deep learning is continuing to increase due to its capacity to improve the accuracy of predictions, enabling improved data-driven decisions, and its ability to learn from unstructured and unlabelled datasets(1). This has led to operational and financial benefits for companies(1).

The number of deep learning publications on arXiv has increased almost six times in the last five years(1). Open source libraries for deep learning like TensorFlow, Keras, PyTorch, Sckit-learn, BVL/caffe, MXNet, and Microsoft Cognitive Toolkit (CNTK) are helping developers easily build deep learning models(1).

Few-shot learning models have potential to strengthen deep learning models with new researches and developments. Generative adversarial networks (GANs) are popular in data augmentation applications and they can create meaningful new data by using unlabelled original data(1).

The current growth of deep learning research and industry applications demonstrate its ubiquitous presence in every facet of AI—be it natural language processing or computer vision applications. With time and research opportunities, unsupervised learning methods may deliver models that will closely mimic human behavior(2).

Source: Conversation with Bing, 2/8/2024

1. Future of Deep Learning according to top AI Experts of 2023—AIMultiple. https://research.aimultiple.com/future-of-deep-learning/.
2. The Future of Deep Learning—DATAVERSITY. https://www.dataversity.net/the-future-of-deep-learning/.
3. The Development of Deep Learning Technologies—Springer. https://link.springer.com/book/10.1007/978-981-15-4584-9.
4. 13 Deep Learning Trends to Watch in 2023 According to Top ...—Deci. https://deci.ai/blog/13-deep-learning-ai-trends-2023/.
5. Getty Images. https://www.gettyimages.com/detail/photo/deep-learning-logo-royalty-free-image/871793108.[6]

[6] Broken link.

References

Here are the references in citation order.

1. Nelson Nazzicari and Filippo Biscarini. 'Stacked kinship CNN vs. GBLUP for genomic predictions of additive and complex continuous phenotypes'. In: *Scientific Reports* 12.1(2022), p. 19889 (cited on pages 2, 129).
2. Stefano Biffani and Filippo Biscarini. 'Deep neural network-based method applied to confocal microscopy images of PDO cheeses'. In: *SensorFINT conference & AK Chemometrik und Qualitätssicherung Annual Workshop*. 2023 (cited on page 2).
3. Chiranjib Chakraborty et al. 'From machine learning to deep learning: An advances of the recent data-driven paradigm shift in medicine and healthcare'. In: *Current Research in Biotechnology* (2023), p. 100164 (cited on page 2).
4. Samuel G. Armato III et al. 'The Lung Image Database Consortium (LIDC) and Image Database Resource Initiative (IDRI): A Completed Reference Database of Lung Nodules on CT Scans'. In: *Medical Physics* 38.2 (2011), pp. 915–931. DOI: https://doi.org/10.1118/1.3528204 (cited on page 3).
5. Arnaud Arindra Adiyoso Setio et al. 'Validation, comparison, and combination of algorithms for automatic detection of pulmonary nodules in computed tomography images: the LUNA16 challenge'. In: *Medical image analysis* 42(2017), pp. 1–13 (cited on page 3).
6. Junji Shiraishi et al. 'Development of a digital image database for chest radiographs with and without a lung nodule: receiver operating characteristic analysis of radiologists' detection of pulmonary nodules'. In: *American journal of roentgenology* 174.1(2000), pp. 71–74 (cited on page 3).
7. Hadrien T Gayap and Moulay A Akhloufi. 'Deep machine learning for medical diagnosis, application to lung cancer detection: a review'. In: *BioMedInformatics* 4.1(2024), pp. 236–284 (cited on page 3).
8. Lulu Wang. 'Deep learning techniques to diagnose lung cancer'. In: *Cancers* 14.22(2022), p. 5569 (cited on page 3).
9. A Asuntha and Andy Srinivasan. 'Deep learning for lung Cancer detection and classification'. In: *Multimedia Tools and Applications* 79.11(2020), pp. 7731–7762 (cited on page 3).
10. Hongming Chen et al. 'The rise of deep learning in drug discovery'. In: *Drug Discovery Today* 23.6(2018), pp. 1241–1250. DOI: https://doi.org/10.1016/j.drudis.2018.01.039 (cited on page 3).
11. Heba Askr et al. 'Deep learning in drug discovery: an integrative review and future challenges'. In: *Artificial Intelligence Review* 56.7(2023), pp. 5975–6037 (cited on page 3).

F. Biscarini, N. Nazzicari, *Deep Learning for Life Sciences*, Decoding Evolution 1,
https://doi.org/10.1007/978-3-031-96852-5

12. Dries Berckmans et al. 'Animal sound. . . talks! Real-time sound analysis for health monitoring in livestock'. In: *Proc. Animal Environment and Welfare*(2015), pp. 215–222 (cited on page 3).
13. Ton J Cleophas and Aeilko H Zwinderman. *Statistics applied to clinical studies*. Springer Science & Business Media, 2012 (cited on page 8).
14. Ben Goertzel and Cassio Pennachin. *Artificial general intelligence*. Vol. 2. Springer, 2007 (cited on page 13).
15. Andreas Christmann and Ingo Steinwart. *Support vector machines*. Springer, 2008 (cited on page 13).
16. David M Blei. 'Probabilistic topic models'. In: *Communications of the ACM* 55.4(2012), pp. 77–84 (cited on page 15).
17. Kari S McLeod. 'Our sense of Snow: the myth of John Snow in medical geography'. In: *Social science & medicine* 50.7-8(2000), pp. 923–935 (cited on page 15).
18. Filippo Biscarini et al. 'Genome-wide association study for traits related to plant and grain morphology, and root architecture in temperate rice accessions'. In: *PloS one* 11.5(2016), e0155425 (cited on page 15).
19. *Data compression and unsupervised learning*. [Online; accessed 02. Dec. 2021]. Feb. 2013. URL: https://lips.cs.princeton.edu/data-compression-and-unsupervised-learning/ (cited on page 15).
20. Adelbert W Bronkhorst. 'The cocktail party phenomenon: A review of research on speech intelligibility in multiple-talker conditions'. In: *Acta Acustica united with Acustica* 86.1(2000), pp. 117–128 (cited on page 15).
21. Friso Kingma, Pieter Abbeel, and Jonathan Ho. 'Bit-swap: Recursive bits-back coding for lossless compression with hierarchical latent variables'. In: *International Conference on Machine Learning*. PMLR. 2019, pp. 3408–3417 (cited on page 15).
22. Sarfaraz Hussein et al. 'Lung and pancreatic tumor characterization in the deep learning era: novel supervised and unsupervised learning approaches'. In: *IEEE transactions on medical imaging* 38.8(2019), pp. 1777–1787 (cited on page 15).
23. Jesper E Van Engelen and Holger H Hoos. 'A survey on semi-supervised learning'. In: *Machine Learning* 109.2(2020), pp. 373–440 (cited on page 16).
24. Shagan Sah. 'Machine learning: a review of learning types'. In: (2020) (cited on page 16).
25. Iqbal H Sarker. 'Machine learning: Algorithms, real-world applications and research directions'. In: *SN computer science* 2.3(2021), p. 160 (cited on page 16).
26. Gareth James et al. *An Introduction to Statistical Learning*. 2013 (cited on page 17).
27. Savi Gupta and Roopal Mamtora. 'A survey on association rule mining in market basket analysis'. In: *International Journal of Information and Computation Technology* 4.4(2014), pp. 409–414 (cited on page 19).
28. Christopher Jermaine. 'Finding the most interesting correlations in a database: How hard can it be?' In: *Information Systems* 30.1 (2005), pp. 21–46 (cited on page 19).
29. Carleton Douglas Creelman. 'Signal Detection Theory, History of'. In: *International Encyclopedia of the Social & Behavioral Sciences(Second Edition)*. Ed. by James D. Wright. Second Edition. Oxford: Elsevier, 2015, pp. 952–956. DOI: https://doi.org/10.1016/B978-0-08-097086-8.43091-6 (cited on page 25).
30. César Ferri, José Hernández-Orallo, and Miguel Angel Salido. 'Volume under the ROC surface for multi-class problems'. In: *European conference on machine learning*. Springer. 2003, pp. 108–120 (cited on page 26).
31. Matthew S Wandishin and Steven J Mullen. 'Multiclass ROC analysis'. In: *Weather and Forecasting* 24.2 (2009), pp. 530–547 (cited on page 26).
32. Jingyan Xu. 'Comparing multi-class classifier performance by multi-class ROC analysis: A nonparametric approach'. In: *Neurocomputing* (2024), p. 127520 (cited on page 26).
33. Quan Zou et al. 'Finding the best classification threshold in imbalanced classification'. In: *Big Data Research* 5 (2016), pp. 2–8 (cited on page 26).
34. Pierre Baldi et al. 'Assessing the accuracy of prediction algorithms for classification: an overview'. In: *Bioinformatics* 16.5 (2000), pp. 412–424 (cited on page 26).

35. Tania Bobbo et al. 'Machine learning classification of archaea and bacteria identifies novel predictive genomic features'. In: *BMC genomics* 25.1 (2024), p. 955 (cited on page 27).
36. Jacob Cohen. 'A coefficient of agreement for nominal scales'. In: *Educational and psychological measurement* 20.1 (1960), pp. 37–46 (cited on page 27).
37. Meelis Kull, Telmo M Silva Filho, and Peter Flach. 'Beyond sigmoids: How to obtain well-calibrated probabilities from binary classifiers with beta calibration'. In:(2017) (cited on page 28).
38. Glenn W Brier. 'Verification of forecasts expressed in terms of probability'. In: *Monthly weather review* 78.1 (1950), pp. 1–3 (cited on page 28).
39. David N Reshef et al. 'Detecting novel associations in large data sets'. In: *science* 334.6062 (2011), pp. 1518–1524 (cited on page 31).
40. C. Spearman. 'The Proof and Measurement of Association between Two Things'. In: *The American Journal of Psychology* 15.1 (1904), pp. 72–101. (Visited on 04/11/2024) (cited on page 31).
41. Kalervo Järvelin and Jaana Kekäläinen. 'Cumulated gain-based evaluation of IR techniques'. In: *ACM Transactions on Information Systems (TOIS)* 20.4 (2002), pp. 422–446 (cited on page 31).
42. Robert Tibshirani. 'Immune signatures in follicular lymphoma'. In: *The New England Journal of Medicine* 352.14 (2005), pp. 1496–1497 (cited on page 33).
43. Gary King and Langche Zeng. 'Logistic regression in rare events data'. In: *Political analysis* 9.2 (2001), pp. 137–163 (cited on page 36).
44. Qiuyu Wang and Henk Bovenhuis. 'Validation strategy can result in an overoptimistic view of the ability of milk infrared spectra to predict methane emission of dairy cattle'. In: *Journal of dairy science* 102.7 (2019), pp. 6288–6295 (cited on page 37).
45. David R Roberts et al. 'Cross-validation strategies for data with temporal, spatial, hierarchical, or phylogenetic structure'. In: *Ecography* 40.8 (2017), pp. 913–929 (cited on page 37).
46. M Pszczola et al. 'Reliability of direct genomic values for animals with different relationships within and to the reference population'. In: *Journal of dairy science* 95.1 (2012), pp. 389–400 (cited on page 38).
47. Michael X Cohen. 'A tutorial on generalized eigendecomposition for denoising, contrast enhancement, and dimension reduction in multichannel electrophysiology'. In: *Neuroimage* (2021), p. 118809 (cited on page 39).
48. Claudia Ottka et al. 'A novel canine nuclear magnetic resonance spectroscopy-based metabolomics platform: validation and sample handling'. In: *Veterinary Clinical Pathology* 50.3 (2021), pp. 410–426 (cited on page 39).
49. David H Alexander, John Novembre, and Kenneth Lange. 'Fast model-based estimation of ancestry in unrelated individuals'. In: *Genome research* 19.9 (2009), pp. 1655–1664 (cited on page 40).
50. Svante Wold. 'Cross-validatory estimation of the number of components in factor and principal components models'. In: *Technometrics* 20.4 (1978), pp. 397-405 (cited on page 40).
51. L Todd Rose and Kurt W Fischer. 'Garbage in, garbage out: Having useful data is everything'. In: *Measurement: Interdisciplinary Research & Perspective* 9.4 (2011), pp. 222–226 (cited on page 41).
52. Enguang Zuo et al. 'Anomaly Score-Based Risk Early Warning System for Rapidly Controlling Food Safety Risk'. In: *Foods* 11.14 (2022), p. 2076 (cited on page 41).
53. Stephan Clémençon and Albert Thomas. 'Mass volume curves and anomaly ranking'. In: *Electronic Journal of Statistics* 12.2 (2018), pp. 2806–2872 (cited on page 41).
54. Richard Baker. 'Lebesgue measure on $\mathbb{R}^\infty$, II'. In: *Proceedings of the American Mathematical Society* 132.9 (2004), pp. 2577–2591 (cited on page 41).
55. Georg Steinbuss and Klemens Böhm. 'Benchmarking unsupervised outlier detection with realistic synthetic data'. In: *ACM Transactions on Knowledge Discovery from Data (TKDD)* 15.4 (2021), pp. 1–20 (cited on page 42).

56. Rob Kitchin. 'Big Data, new epistemologies and paradigm shifts'. In: *Big data & society* 1.1 (2014), p. 2053951714528481 (cited on page 42).
57. Bradley Efron. 'Bootstrap methods: another look at the jackknife'. In: *Breakthroughs in statistics: Methodology and distribution*. Springer, 1992, pp. 569–593 (cited on page 43).
58. François Pompanon et al. 'Genotyping errors: causes, consequences and solutions'. In: *Nature Reviews Genetics* 6.11 (2005), pp. 847–859 (cited on page 44).
59. Steven Buyske et al. 'When a case is not a case: effects of phenotype misclassification on power and sample size requirements for the transmission disequilibrium test with affected child trios'. In: *Human heredity* 67.4 (2009), pp. 287–292 (cited on page 44).
60. Stefano Biffani et al. 'The effect of mislabeled phenotypic status on the identification of mutation-carriers from SNP genotypes in dairy cattle'. In: *BMC research notes* 10 (2017), pp. 1–9 (cited on page 44).
61. Filippo Biscarini et al. ' "Noisy beets": impact of phenotyping errors on genomic predictions for binary traits in Beta vulgaris'. In: *Plant Methods* 12 (2016), pp. 1–8 (cited on page 44).
62. Mark PJ Van der Loo. *Distribution based outlier detection in univariate data*. Statistics Netherlands, 2010 (cited on page 44).
63. Rob J Hyndamn. *That's weird! Anomaly detection using R*. [Online; accessed 18. Apr. 2024]. Monash University, Australia, Mar. 2024. URL: https://otexts.com/weird/ (cited on page 44).
64. Michał Oleksowicz. 'Aristotle on the Heart and Brain'. In: *European Journal of Science and Theology* 14.3 (2018), pp. 77–94 (cited on page 47).
65. Warren S McCulloch and Walter Pitts. 'A logical calculus of the ideas immanent in nervous activity'. In: *The bulletin of mathematical biophysics* 5.4 (1943), pp. 115–133 (cited on page 47).
66. Donald Olding Hebb. *The organization of behavior: A neuropsychological theory*. Psychology Press, 2005 (cited on page 48).
67. Frank Rosenblatt. *The perceptron, a perceiving and recognizing automaton (Project Para)*. Cornell Aeronautical Laboratory, 1957 (cited on page 48).
68. Marvin Minsky and Seymour Papert. 'Perceptrons: an introduction to computational geometry, expanded edition'. In: *Editorial The MIT Pres. Libro publicado* 28 (1987) (cited on page 48).
69. Frank Rosenblatt. *Principles of neurodynamics. perceptrons and the theory of brain mechanisms*. Tech. rep. Cornell Aeronautical Lab Inc Buffalo NY, 1961 (cited on page 48).
70. Bruce G Buchanan and Reid G Smith. 'Fundamentals of expert systems'. In: *Annual review of computer science* 3.1 (1988), pp. 23–58 (cited on page 48).
71. David E Rumelhart, Geoffrey E Hinton, and Ronald J Williams. 'Learning representations by back propagating errors'. In: *nature* 323.6088 (1986), pp. 533–536 (cited on page 48).
72. Julian G Barwell et al. 'Challenges in implementing genomic medicine: the 100,000 Genomes Project'. In: *Journal of Translational Genetics and Genomics* 2 (2018), p. 13 (cited on page 49).
73. Hehuan Ma, Yu Rong, and Junzhou Huang. 'Graph Neural Networks: Scalability'. In: *Graph Neural Networks: Foundations, Frontiers, and Applications*. Springer, 2022, pp. 99–119 (cited on page 51).
74. Kurt Hornik, Maxwell Stinchcombe, and Halbert White. 'Multilayer feedforward networks are universal approximators'. In: *Neural networks* 2.5 (1989), pp. 359–366 (cited on page 51).
75. Xavier Glorot, Antoine Bordes, and Yoshua Bengio. 'Deep sparse rectifier neural networks'. In: *Proceedings of the fourteenth international conference on artificial intelligence and statistics*. JMLR Workshop and Conference Proceedings. 2011, pp. 315–323 (cited on page 56).
76. Andrew L Maas, Awni Y Hannun, Andrew Y Ng, et al. 'Rectifier nonlinearities improve neural network acoustic models'. In: *Proc. icml*. Vol.30.1. Atlanta, Georgia, USA. 2013, p. 3 (cited on page 57).
77. Meenal V Narkhede, Prashant P Bartakke, and Mukul S Sutaone. 'A review on weight initialization strategies for neural networks'. In: *Artificial intelligence review* 55.1 (2022), pp. 291–322 (cited on page 63).

78. Ronald A Fisher. 'The use of multiple measurements in taxonomic problems'. In: *Annals of eugenics* 7.2(1936), pp. 179–188 (cited on page 67).
79. Pavel Tokmakov, Yu-Xiong Wang, and Martial Hebert. 'Learning compositional representations for few-shot recognition'. In: *Proceedings of the IEEE/CVF International Conference on Computer Vision*. 2019, pp. 6372–6381 (cited on page 69).
80. Thao Nguyen, Maithra Raghu, and Simon Kornblith. 'Do wide and deep networks learn the same things? uncovering how neural network representations vary with width and depth'. In: *arXiv preprint arXiv:2010.15327* (2020) (cited on page 70).
81. Gal Vardi, Gilad Yehudai, and Ohad Shamir. 'Width is less important than depth in relu neural networks'. In: *Conference on Learning Theory*. PMLR. 2022, pp. 1249–1281 (cited on page 70).
82. Adityanarayanan Radhakrishnan, Mikhail Belkin, and Caroline Uhler. 'Wide and deep neural networks achieve consistency for classification'. In: *Proceedings of the National Academy of Sciences* 120.14(2023), e2208779120 (cited on page 70).
83. Sergey Ioffe and Christian Szegedy. 'Batch normalization: Accelerating deep network training by reducing internal covariate shift'. In: *International conference on machine learning*. pmlr. 2015, pp. 448–456 (cited on page 75).
84. Jimmy Lei Ba, Jamie Ryan Kiros, and Geoffrey E Hinton. 'Layer normalization'. In: *arXiv preprint arXiv:1607.06450*(2016) (cited on page 76).
85. Kaiming He et al. 'Deep residual learning for image recognition'. In: *Proceedings of the IEEE conference on computer vision and pattern recognition*. 2016, pp. 770–778 (cited on page 76).
86. Fengxiang He, Tongliang Liu, and Dacheng Tao. 'Why resnet works? residuals generalize'. In: *IEEE transactions on neural networks and learning systems* 31.12 (2020), pp. 5349–5362 (cited on page 76).
87. Hyeonwoo Noh et al. 'Regularizing deep neural networks by noise: Its interpretation and optimization'. In: *Advances in neural information processing systems* 30 (2017) (cited on page 83).
88. Nitish Srivastava et al. 'Dropout: a simple way to prevent neural networks from overfitting'. In: *The journal of machine learning research* 15.1 (2014), pp. 1929–1958 (cited on page 84).
89. Marc Brysbaert et al. 'How many words do we know? Practical estimates of vocabulary size dependent on word definition, the degree of language input and the participant's age'. In: *Frontiers in psychology* 7 (2016), p. 190735 (cited on page 96).
90. Douwe Kiela and Léon Bottou. 'Learning image embeddings using convolutional neural networks for improved multi-modal semantics'. In: *Proceedings of the 2014 Conference on empirical methods in natural language processing (EMNLP)*. 2014, pp. 36–45 (cited on page 96).
91. Sepp Hochreiter and Jürgen Schmidhuber. 'Long short-term memory'. In: *Neural computation* 9.8 (1997), pp.1735–1780 (cited on page 100).
92. Kyunghyun Cho et al. 'On the properties of neural machine translation: Encoder-decoder approaches'. In: *arXiv preprint arXiv:1409.1259* (2014) (cited on page 100).
93. Junyoung Chung et al. 'Empirical evaluation of gated recurrent neural networks on sequence modeling'. In: *arXiv preprint arXiv:1412.3555* (2014) (cited on page 101).
94. George EP Box and David R Cox. 'An analysis of transformations'. In: *Journal of the Royal Statistical Society Series B: Statistical Methodology* 26.2 (1964), pp. 211–243 (cited on page 107).
95. Edward L Fink. 'The FAQs on data transformation'. In: *Communication Monographs* 76.4 (2009), pp. 379–397 (cited on page 107).
96. Clemens Reimann et al. *Statistical data analysis explained: applied environmental statistics with R*. John Wiley & Sons, 2011 (cited on page 107).
97. Azra Ramezankhani et al. 'The impact of oversampling with SMOTE on the performance of 3 classifiers in prediction of type 2 diabetes'. In: *Medical decision making* 36.1 (2016), pp. 137–144 (cited on page 108).
98. Giovanna Menardi and Nicola Torelli. 'Training and assessing classification rules with imbalanced data'. In: *Data mining and knowledge discovery* 28 (2014), pp. 92–122 (cited on page 108).

99. Olaf Ronneberger, Philipp Fischer, and Thomas Brox. 'U-net: Convolutional networks for biomedical image segmentation'. In: *Medical Image Computing and Computer-Assisted Intervention–MICCAI 2015: 18th International Conference, Munich, Germany, October 5-9, 2015, Proceedings, Part III 18*. Springer. 2015, pp. 234–241 (cited on page 114).
100. Alexander Kirillov et al. 'Segment anything'. In: *Proceedings of the IEEE/CVF International Conference on Computer Vision*. 2023, pp. 4015–4026 (cited on page 115).
101. Kaiming He et al. 'Mask r-cnn'. In: *Proceedings of the IEEE international conference on computer vision*. 2017, pp. 2961–2969 (cited on page 116).
102. Ross Girshick. 'Fast r-cnn'. In: *Proceedings of the IEEE international conference on computer vision*. 2015, pp. 1440–1448 (cited on page 116).
103. Nicolas Carion et al. 'End-to-end object detection with transformers'. In: *European conference on computer vision*. Springer. 2020, pp. 213–229 (cited on page 117).
104. Castrense Savojardo, Pier Luigi Martelli, and Rita Casadio. 'Finding functional motifs in protein sequences with deep learning and natural language models'. In: *Current Opinion in Structural Biology* 81 (2023), p. 102641 (cited on page 118).
105. Akosua Busia et al. 'A deep learning approach to pattern recognition for short DNA sequences'. In: *BioRxiv* (2018), p. 353474 (cited on page 118).
106. Smith K Khare et al. 'Emotion recognition and artificial intelligence: A systematic review (2014-2023) and research recommendations'. In: *Information Fusion* (2023), p. 102019 (cited on page 118).
107. Ayham Alomari et al. 'Deep reinforcement and transfer learning for abstractive text summarization: A review'. In: *Computer Speech & Language* 71 (2022), p. 101276 (cited on page 118).
108. Deepak Suresh Asudani, Naresh Kumar Nagwani, and Pradeep Singh. 'Impact of word embedding models on text analytics in deep learning environment: a review'. In: *Artificial intelligence review* 56.9 (2023), pp. 10345–10425 (cited on page 118).
109. Martin Popel et al. 'Transforming machine translation: a deep learning system reaches news translation quality comparable to human professionals'. In: *Nature communications* 11.1 (2020), pp. 1–15 (cited on page 118).
110. S Bhattacharya et al. 'Deep classification of sound: A concise review'. In: *Proceeding of First Doctoral Symposium on Natural Computing Research: DSNCR 2020*. Springer. 2021, pp. 33–43 (cited on page 118).
111. Yoonjoo Kim et al. 'The coming era of a new auscultation system for analyzing respiratory sounds'. In: *BMC Pulmonary Medicine* 22.1 (2022), p. 119 (cited on page 119).
112. Eddiemar B Lagua et al. 'A Comprehensive Monitoring System for Health Status and Behavior of Pigs Under Different Environmental Conditions Using Vision and Audio Technologies'. In: (2023) (cited on page 119).
113. Yihan Cao et al. 'A comprehensive survey of ai-generated content (aigc): A history of generative ai from gan to chatgpt'. In: *arXiv preprint arXiv:2303.04226* (2023) (cited on page 119).
114. Keras Team. *Keras Documentation: Keras applications*. URL: https://keras.io/api/applications/ (cited on page 120).
115. Kaggle Team. *Repository of pre-trained models*. URL: https://www.kaggle.com/models (cited on page 120).
116. Michael TM Emmerich and André H Deutz. 'A tutorial on multiobjective optimization: fundamentals and evolutionary methods'. In: *Natural computing* 17 (2018), pp. 585–609 (cited on page 122).
117. Nyoman Gunantara. 'A review of multi-objective optimization: Methods and its applications'. In: *Cogent Engineering* 5.1 (2018), p. 1502242 (cited on page 122).
118. R Timothy Marler and Jasbir S Arora. 'Survey of multi-objective optimization methods for engineering'. In: *Structural and multidisciplinary optimization* 26 (2004), pp. 369–395 (cited on page 122).
119. Diederik P Kingma and Jimmy Ba. 'Adam: A method for stochastic optimization'. In: *arXiv preprint arXiv:1412.6980* (2014) (cited on page 125).

120. Geoffrey Hinton, Nitish Srivastava, and Kevin Swersky. 'Neural networks for machine learning lecture 6a overview of mini-batch gradient descent'. In: *Cited on* 14.8 (2012), p. 2 (cited on page 125).
121. Léon Bottou. 'Stochastic gradient descent tricks'. In: *Neural Networks: Tricks of the Trade: Second Edition* (2012), pp. 421–436 (cited on page 125).
122. Dheeru Dua and Casey Graff. *UCI Machine Learning Repository*. 2017. URL:http://archive.ics.uci. edu/ml (cited on page 126).
123. Gabrielle Ras et al. 'Explainable deep learning: A field guide for the uninitiated'. In: *Journal of Artificial Intelligence Research* 73 (2022), pp. 329–396 (cited on page 129).
124. Varun Godbole et al. *Deep Learning Tuning Playbook*. Version 1.0.2023. URL: http://github.com/ google-research/tuning_playbook (cited on page 130).
125. Xi Yang et al. 'Gatortron: A large clinical language model to unlock patient information from unstructured electronic health records'. In: *arXiv preprint arXiv:2203.03540* (2022) (cited on page 137).
126. Ian J Goodfellow, Jonathon Shlens, and Christian Szegedy. 'Explaining and harnessing adversarial examples'. In: *arXiv preprint arXiv:1412.6572* (2014) (cited on page 139).
127. Jakob Gawlikowski et al. 'A survey of uncertainty in deep neural networks'. In: *Artificial Intelligence Review* 56.Suppl 1 (2023), pp. 1513–1589 (cited on page 139).
128. Murat Seckin Ayhan and Philipp Berens. 'Test-time data augmentation for estimation of heteroscedastic aleatoric uncertainty in deep neural networks'. In: *Medical Imaging with Deep Learning*. 2022 (cited on page 139).
129. Ashish Vaswani et al. 'Attention is all you need'. In: *Advances in neural information processing systems* 30 (2017) (cited on page 146).
130. Dzmitry Bahdanau, Kyunghyun Cho, and Yoshua Bengio. 'Neural machine translation by jointly learning to align and translate'. In: *arXiv preprint arXiv:1409.0473* (2014) (cited on page 146).
131. Alex Hernandez-Garcia et al. 'Multi-fidelity active learning with GFlowNets'. In: *arXiv preprint arXiv:2306.11715* (2023) (cited on page 149).
132. Pengzhen Ren et al. 'A survey of deep active learning'. In: *ACM computing surveys (CSUR)* 54.9 (2021), pp. 1–40 (cited on page 153).
133. Vu-Linh Nguyen, Mohammad Hossein Shaker, and Eyke Hüllermeier. 'How to measure uncertainty in uncertainty sampling for active learning'. In: *Machine Learning* 111.1 (2022), pp. 89–122 (cited on page 154).
134. Sandra Gilhuber et al. 'Accelerating Diversity Sampling for Deep Active Learning By Low-Dimensional Representations.' In: *IAL@PKDD/ECML*. 2022, pp. 43–48 (cited on page 154).
135. Hideitsu Hino and Shinto Eguchi. 'Active learning by query by committee with robust divergences'. In: *Information Geometry* 6.1 (2023), pp. 81–106 (cited on page 154).
136. Yann LeCun, Yoshua Bengio, et al. 'Convolutional networks for images, speech, and time series'. In: *The handbook of brain theory and neural networks* 3361.10 (1995), p. 1995 (cited on page 155).
137. Jerome Kelleher, Alison M Etheridge, and Gilean McVean. 'Efficient coalescent simulation and genealogical analysis for large sample sizes'. In: *PLoS computational biology* 12.5 (2016), e1004842 (cited on page 157).
138. Justin T Reese et al. 'KG-COVID-19: a framework to produce customized knowledge graphs for COVID-19 response'. In: *Patterns* 2.1 (2021) (cited on page 158).
139. Michael M Bronstein et al. 'Geometric deep learning: going beyond euclidean data'. In: *IEEE Signal Processing Magazine* 34.4 (2017), pp. 18–42 (cited on page 161).
140. Jie Zhou et al. 'Graph neural networks: A review of methods and applications'. In: *AI open* 1 (2020), pp. 57–81 (cited on page 161).
141. Zonghan Wu et al. 'A comprehensive survey on graph neural networks'. In: *IEEE transactions on neural networks and learning systems* 32.1 (2020), pp. 4–24 (cited on page 161).
142. F. Wong, E.J. Zheng, and J.A. et al Valeri. 'Discovery of a structural class of antibiotics with explainable deep learning'. In: *Nature* (2023). DOI: https://doi.org/10.1038/s41586-023-06887-8 (cited on page 161).